Francis Asbury Shoup

Mechanism and Personality

An Outline of Philosophy in the Light of the Latest Scientific Research

Francis Asbury Shoup

Mechanism and Personality
An Outline of Philosophy in the Light of the Latest Scientific Research

ISBN/EAN: 9783337269869

Printed in Europe, USA, Canada, Australia, Japan

Cover: Foto ©berggeist007 / pixelio.de

More available books at **www.hansebooks.com**

MECHANISM AND PERSONALITY:

AN

OUTLINE OF PHILOSOPHY IN THE LIGHT OF THE LATEST SCIENTIFIC RESEARCH.

BY

FRANCIS A. SHOUP, D.D.,

PROFESSOR OF ANALYTICAL PHYSICS, UNIVERSITY OF THE SOUTH.

BOSTON, U.S.A.:
PUBLISHED BY GINN & COMPANY.
1891.

TYPOGRAPHY BY J. S. CUSHING & CO., BOSTON, U.S.A.
PRESSWORK BY GINN & CO., BOSTON, U.S.A.

To
My Friend,
THOMAS BRESLIN,
who,
of all the men it has been my happiness
to know, is the most thoroughly
altruistic.

> "The truth which draws
> Through all things upwards; that a two-fold world
> Must go to a perfect cosmos. Natural things
> And spiritual, — who separates those two
> In art, in morals, or the social drift,
> Tears up the bond of nature and brings death,
> Paints futile pictures, writes unreal verse,
> Leads vulgar days, deals ignorantly with men,
> Is wrong, in short, at all points."
> — *Aurora Leigh.*

"The reconciliation of physics and metaphysics lies in the acknowledgment of faults upon both sides; in the confession by physics that all the phenomena of nature are, in their ultimate analysis, known to us only as facts of consciousness ; in the admission by metaphysics, that the facts of consciousness are, practically, interpretable only by the methods and the formulæ of physics." — *Professor Huxley.*

PREFACE.

SOME time ago a gentleman of excellent attainments requested the author of the following pages to recommend him a book which would give, within moderate compass, the present attitude of Philosophy in the light of the latest scientific research, and that in a way suited to the comprehension of the ordinary reader. The author mentioned several books which he thought would in some sort answer the purpose, but at the same time had to confess that he could think of no one work which exactly met the case. Reflecting afterwards from time to time upon the subject, he was still unable to fix upon any one such book.

It was in this way that the need of something to meet the growing inquiry as to what has become of metaphysic in the glare of the scientific thought of the day impressed itself upon the author, and that he conceived the idea of trying what he could do himself in the way of outlining an answer. These pages are the result of his effort.

The author has tried to keep the general reader in mind, and, as a result, the book is largely elementary; but, while aiming at simplicity and clearness, he has not thought it best to avoid entirely the recognized terminology of the subject. Care has been taken, however, when introducing purely technical terms, to give equivalent expressions in common speech. The author has been at times tempted — almost compelled —

to enter upon disputed ground, and to venture upon questions of considerable subtlety, but, for all that, the book will be found, in the main, fairly easy reading.

With regard to materials, a free hand has been laid upon whatever was within reach; and although it has been thought unnecessary to give detailed references, the reader will be able to know, in a general way, to whom credit is due. The author is free to confess his regret, however, that his references have been so meagre.

The reader will find that there has been no effort to keep back or underrate the conclusions of the most advanced scientific thought, but that the burning questions between the Empiricists and Transcendentalists have been treated with perfect candor and openness.

The metaphysic is, in the main, that of Lotze, or perhaps better, the Lotzian phase of Kant. The "Outlines of Metaphysic," lately translated by Professor Ladd (Ginn & Company), has been found most suitable for quotation, and, with the exception of a few passages from the "Mikrokosmus," the extracts from Lotze may be found in that English version.

The thanks of the author are due to Dr. Henry H. Donaldson, Professor of Neurology in Clark University, for valuable assistance in revising Chapters III. to VII. inclusive, and to John Fearnley, M.A., of the University of the South, for efficient help in revising proofs.

If this book shall be the means of directing speculative thought more in the lines marked out by Lotze, it will have served a useful purpose, in the opinion of

THE AUTHOR.

SEWANEE, TENN., January, 1891.

CONTENTS.

CHAPTER I.

SCOPE AND LIMITS OF SKEPTICISM 1

What is truth? Incompleteness of Knowledge. Lack of permanence. Difficulties of arriving at ultimate principles. The senses mendacious. The intellect open to attack. May we not be compelled to see things as they are not? Practical limits of doubt. Logical limits. Personality. The self an ultimate fact. The 'One and the many.'

CHAPTER II.

THE MECHANICAL BASIS OF PHENOMENA 11

Modern science and the older learning. All science dominated by mechanics. Reducible to mathematical forms. Forestalled by Descartes. Hobbes. Leibnitz. Attitude of modern physicists. Metaphysical basis of science obvious to all thinkers.

CHAPTER III.

PSYCHO-MECHANISMS 18

The cell-theory discarded. Protoplasmic movement. Max Schultze. Huxley. Uni-cellular organisms. Structural development. Professor Foster quoted. 'Metabolism.' Nervous system. Reflex action. Vivisection. Cerebral hemispheres. Effects of mutilations. Caution. Functions of different brain-areas not certainly determined.

CHAPTER IV.

PSYCHO-MECHANISMS (*continued*) 28

Professor Romanes quoted. Experiments in vivisection. Brain localizations. Electrical stimulations. Rate of nerve transmissions. Time required for action of nerve-centres. Rate of nerve-vibrations. The sympathetic system. Functions. Independent of volition. Inhibition. Brain-development. Brain-mass. Different nationalities.

CHAPTER V.

THE SENSES — TOUCH, TASTE, AND SMELL 38

The specific senses. Number indefinite. Touch fundamental. Pressure. End-organs. Threshold value. Weber's and Fechner's law. 'Local signs.' Pressure spots. Temperature spots. End-organs of taste. Stimuli. Classification. Sense of smell. End-organs of smell. Stimuli. Classification. Muscular sense. End-organs of motion.

CHAPTER VI.

THE SENSE OF HEARING 48

The ear. Structure. Corti's organ. Theories. Physical basis of sound. Intensity, pitch, quality. Illustrations. Partials. Tyndall quoted. Difference in people's sensibilities. Powers of discrimination. Range. The human voice.

CHAPTER VII.

THE SENSE OF VISION 60

Mechanism of the eye. Structure of the retina. End-organs. Rods and cones. Mechanical basis of vision. Color. 'Consecutive' images. Tone, intensity, saturation. Yellow spot.

CHAPTER VIII.

CHASM BETWEEN MECHANISM AND CONSCIOUSNESS 69

Physiological research with respect to psycho-mechanisms. Protoplasm not pure and simple matter. Professor Romanes quoted.

Darwin, Huxley, Tyndall, and Spencer not materialists. Hobbes quoted. The problem of relation between physiology and consciousness. The chasm recognized. Leaders of science quoted.

CHAPTER IX.

PERSONALITY IN ITS PSYCHICAL ASPECT 82

Analysis of the psychical factor of personality. The three fundamental modes of the self—sensation, cognition, and conation. A tri-unity, inseparable but logically distinguishable. Sub-consciousness. Unity and plurality.

CHAPTER X.

DEVELOPMENT OF THE PSYCHICAL ASPECT OF PERSONALITY . . . 86

The relation of the mechanical and psychical factors. Mutually necessary. The human organism at birth. The line between elementary consciousness and self-realization shadowy. Automatic action. Basic-personality. Evolution. Continuity and discontinuity. Instincts. 'Jelly-specks.' Ants. *Chætodon rostratus.* The beaver. Domestic animals. Inverse order of intelligence and instinct. A *de*volution as well as an evolution. Instincts gradually replaced in ascending order of nature.

CHAPTER XI.

THE CONCEPT-FORMING PROCESS 97

Muscular co-ordination. Education of the organism. Vital organs not under control of will. Analogous psychical conditions. Process of thought-development. Like and unlike. Discovery of meaning. Attention. Retention. Concepts. Concept-masses. Apperception. Thought as thought. Language. Introspection. 'Pure' and 'empirical ego.' 'One' and 'many.' A sense of 'knowing' deeper than understanding. Personality antedates knowledge.

CHAPTER XII.

DIFFERENTIATIONS OF SELF. MEMORY 109

Consciousness. Differentiation of feeling, of cognition, of will. An end ideally first. Self-development. Perception. Intuition. Ideas in the mind not *like* objects without. Space. Time. Memory. Mechanical basis. Objection by Lotze. Complexity. Illustration from sound. Phenomena explicable upon theory of mechanical basis. Dr. Rush's case. Dr. Carpenter's Welshman. Coleridge's case. Power to recall the past. Sudden recollections. Law of association.

CHAPTER XIII.

THE IMAGINATION 126

Definition. Classification. Cognitive and Sentient imagination. Economic and Rational. Artistic and Rhythmic. Music. Relation of memory and imagination.

CHAPTER XIV.

DREAMING. SOMNAMBULISM. HYPNOTISM 133

Phenomenon of dreaming. Sleep. Do we always dream in sleep? The brain a thought-machine. Consciousness a mere phenomenon. The brain in sleep. Mosso's observations. Character of dreams. Nightmare. Somnambulism. Case of student at Amsterdam. Case recorded by Dr. Abercrombie. German monk. Muscular feats. Double consciousness. Case of young lady at West Point. Hypnotism. Muscular effects. Dr. Charcot quoted. 'Suggestions.'

CHAPTER XV.

THE UNDERSTANDING 149

A technical phase of cognition. Faculty of Relations. Thought proper. The lower animals. Pain. The logical element in man. The syllogism. Dictum of Aristotle. Deduction and Induction. Reasoning. Reciprocal processes.

CONTENTS.

CHAPTER XVI.

THE PURE REASON 161

Hypotheticals. Intuitive Knowledge. Conditions of all explicit thought. Controversy about 'Innate Ideas.' Empirical Knowledge. Law of Identity. Law of Contradiction. Excluded Middle. Its questionable use in certain cases. Hamilton. Sufficient Reason. Causality. Hume. Locke. Leibnitz. The Laws of Motion. All Science based upon necessary truths.

CHAPTER XVII.

EMPIRICAL AND RATIONAL TRUTH 175

Conditional syllogisms and law of sufficient Reason. No law of the natural world above doubt. Not so in thought. '*A priori*,' 'original,' etc., truths. Necessity the characteristic. Relation of Induction and Deduction. The basis of Induction. Intuition of space. The Infinite and Absolute. The 'Philosophy of the Conditioned.'

CHAPTER XVIII.

THE BEARING OF EMPIRICISM ON PERSONALITY 185

Intuition of Time. Time the ground of motion. Space of mass. Cause conditions Space and Time. Inertia. Self-activity inconceivable in 'Thing.' Personality the only ground of efficient cause. 'Persistent Force.' Doubt as to the being of 'force' as an entity. Professor Tait quoted. Spencer's effort to find an ultimate Reality. Energy implies Personality. Spencer's position sounder than that of his followers. Quoted. His 'unknown' known.

CHAPTER XIX.

FEELING 199

Classification. Pain and pleasure. Sensuous feeling. Herbartian scheme. Intensity and quality in feeling. Cænesthesia. Esoteric and exoteric feeling. The one working from within emerges in the understanding; the other built up through the understanding. Practical bearing.

CHAPTER XX.

FEELING (*continued*) 211

    Rational Feeling. Esthetic Feeling. Beauty. Periodic motion. Music. Vision. Illusions. Berkeley's Theory of Vision. Knowledge through vision. Cheselden's case. Other cases. Other problems.

CHAPTER XXI.

FEELING (*continued*) 220

    Art. Ideal element. Sculpture. Painting. Music. Architecture. Poetry. Evolution of ethical feeling. The Good. Ethical treatment reserved to later stage.

CHAPTER XXII.

THE WILL 229

    Elementary effort. Emerges in conscious volition. Much that is commonly accounted free, mechanical. Liberty restricted to Purposive epoch. Inhibitory functions. Directive functions. The office of the will in developing emotional nature. Development of volitive powers. Moral aspect of the will. Penitence.

CHAPTER XXIII.

UNITY OF PERSONALITY 242

    Difficulties of question. Unity and manifoldness. Unity a primordial condition. Inferior organisms. Protozoa. Not two worlds, one spiritual and the other physical. Man a manifestation of one person in two modes. The psychical and mechanical inseparable. Gross and sublimated matter. Visible and Invisible Universe. The 'Unseen Universe' quoted. The mechanical mode has its only title to reality through personality. The Cicada.

CHAPTER XXIV.

WHAT IS 'THING'? CONSTRUCTION OF MATTER 254

    Illusions. What underlies phenomena? Pure Being. 'Thing' that which affects and is affected. The position of Bishop Berkeley. Quoted. Analytical physics and construction of matter. Boscovich's theory. Molecular mechanics. Clerk Maxwell. Professor Tait. Sir W. Thompson's 'vortical atom.' Difficulties. Le Sage's theory. Ether. The physicists driven into metaphysics Atoms 'manufactured articles.'

CHAPTER XXV.

MATHEMATICS NOT ULTIMATELY EXACT 275

Position of mathematics in scientific inquiries. Mathematical processes develop contradictions. Surds. Asymptotes. Graphical illustration. Cissoid of Diocles. Other cases. Right lines intersecting with no common point. The concept 'infinity.' Illustration.

CHAPTER XXVI.

THE METAPHYSICAL ATTITUDE OF CHANGE. CAUSE 283

The problem of change. Quotation from Plato. The problem of causation. Influence 'passing over.' Doctrine of 'Occasionalism.' 'Pre-established Harmony.' 'Divine Assistance.' Lotze quoted. Causation, as such, inexplicable.

CHAPTER XXVII.

RELATION OF PERSONALITY TO SPACE AND TIME, MASS AND MOTION . 294

The concepts Space and Time. Subjective ground of Mass and Motion. Not self-subsisting realities. Find their reality in Personality. Reality of the Cosmos Personal. Soundness of scientific methods. No truth material. Personality necessary to Truth. Personality not a phenomenon.

CHAPTER XXVIII.

SOME OF THE GREAT METAPHYSICAL SYSTEMS 300

Idealism. Fichte. Lotze quoted. Schelling. Hegelianism. Hegel quoted. Objections to absolute Idealism. Lotze's position commended. The Supreme Good.

CHAPTER XXIX.

ETHICAL 313

Self-activity necessary to morality. No liberty in Sensibility or Cognition as such. Choice. Motives. The 'Good.' Obligation. Man held to be omniscient. No obligation in Selfness. Altruism. How the will of the Supreme Good is Known. The 'Categorical Imperative.'

CHAPTER XXX.

THE NATURE AND FUNCTIONS OF CONSCIENCE 325

The admonitions of the moral monitor. Conscience discovers itself only upon change of moral purpose. Analogy between the functions of conscience and inertia. Analysis. Illustration of steamship. Moral momentum.

CHAPTER XXXI.

THE INFINITE PERSONALITY 333

Personal good implies Personality in God. The Mosaic account of the origin of evil in man. Disobedience. Obedience. A class of theologians faulted. Conflict and agreement of the Finite and Infinite. Theology and Religion. Human aspirations. Quotation from Mrs. Browning. Conclusion.

MECHANISM AND PERSONALITY.

CHAPTER I.

SCOPE AND LIMITS OF SKEPTICISM.

What is Truth? Incompleteness of knowledge. Lack of permanence. Difficulties in arriving at ultimate principles. The senses mendacious. The intellect open to attack. May we not be compelled to see things as they are not? Practical limits of doubt. Logical limits. Personality. The Self an ultimate fact. 'The one and the many.'

THE lack of certitude in human knowledge and human destiny has always been a ground of anxiety and complaint. It is not alone a Roman Procurator, who, in mockery or despair, asks the perplexing question, "What is Truth?"; but every man, who thinks at all, must find the inquiry forcing itself upon his attention at times; and, too often, with the dismal refrain, "Who shall show us any good?" And yet it is a great mistake to inveigh against the skeptical element in man's nature. A little reflection must make it apparent that doubt is a necessary condition of knowledge in any form. Paradoxically stated, if we could know at once and perfectly, we should not know at all. It is not by light alone that objects are seen in the external world. If all shade and shadow were removed, there might be left indeed a dead, perfectly even and unvarying illumination; but all sense of sight would be gone, and the external world blotted out. Doubt is the shadow of certitude, without which the world of Truth would vanish from human consciousness.

It must be confessed that the extent to which skepticism may be carried is bewildering in the extreme; but there is a limit which must be reached in thought at last — a practical limit unconsciously forced upon the unthinking at all times, and from which the tide turns back in a flood of unmistakable verity. It will be well to take a look into this yawning chasm of doubt and denial, that we may be more sure of our footing on the heights of the true and abiding.

In the first place, every body knows how incomplete at best is all human knowledge. Those who know the most are most impressed with a sense of their inadequacy and ignorance. Every school-boy knows how Socrates could not understand the declaration of the oracle that he was the wisest man in Athens, until it came over him that it must be because he knew his own ignorance; and how Sir Isaac Newton could fancy himself a little child gathering shells cast up by the sea, while the great ocean of Truth lay spread out in mystery before him. An 'educated ignorance' is but another expression for the highest stage of human knowledge.

Then again the little that one does know has no fixedness and permanence. Every day brings changes — every varying mood causes modifications and colorings. The most firmly settled opinions are constantly assailed by suggestions of doubt. No statement — even the simplest — can be made that is not open to question. Suppose one should attempt to teach a child something of the law of gravitation. He lets a pebble drop from his hand, and asks the child why it falls, explaining that it is because all bodies are attracted to the centre of the earth. The explanation is not true. That body and all other bodies are attracted, not by the earth alone, but by the sun and moon and all other bodies. He corrects himself, enunciating the law of gravitation: Every body attracts every other body with a force which varies inversely as the square of the distance, and directly as the mass. False again. This is only true for sensible distances. When one passes within molecular limits, the law undergoes a change, — perhaps a thousand alternations of attraction and repulsion; and it is hopeless to attempt to follow it. But one cannot be allowed to go even so far in peace. What is a body? No one knows. We may say it is that which is made up of particles, which are composed of molecules, which in turn are composed of atoms. But what is an atom? We do not know. What is a molecule? It is an hypothetical combination or system of atoms, — that is, of elements which nobody can know nor conceive of. Are there any atoms? Nobody knows. It is not only doubted, but stoutly denied. What is attraction or repulsion? Nobody knows. What is force? Not only does nobody know; but the advanced mathematicians and physicists are so seriously skeptical as to the existence of any such 'thing' that they are doing their best to banish the word from the vocabulary of scientific terms. Thus there is not much left of the definition of gravitation; but even what remains is equally open to doubt.

It must be confessed that the skeptic has got on pretty well already in his work of demolition; but he has broad fields yet before him for the exercise of his destructive propensity. He attacks the entire external world, and denies its existence. Take any object, as the long-suffering tree of the metaphysicians: how do we know that it exists? We see it. See what? The color — light and shade. But are these the tree? No; they are the sensations produced through the eye by modifications of light. Then they are something which the tree effects; not what it is.

But it will be said, We can touch it. Yes, and what do we find? That it is hard, rough, cold and all that: — but these are not the tree. They are states or conditions, — called 'properties' or 'accidents' — and so we may go on through every possible test of the senses, and we shall only find other properties, or accidents, no one of which, nor all of them combined, can be the tree. The skeptic still asks, what is 'it'— what is 'The-thing-in-itself?'

He asks further, When one sees objects in a dream, are they real? Have they 'thingness' which supports the properties or attributes, such as color, shape, size and whatever else they seem to have? The sleeper thinks so at the time. Why may it not be that we only *think* they have all these in our waking moments? Or, how do we know which are our waking moments? Why may not the dream-world be the reality, and that which we call the real world be the dream?

But even yet the skeptic is not satisfied. He has cast suspicion upon the external world; he next attacks the thought-world.

One must get all material of thought through experience, and experience must come through the senses. But the senses are not infallible. By an artful combination of mirrors, the most successful delusions are practised by the modern magician. Every body is misled from time to time by tricks of vision. The ear is even more fallible, and all the senses, even touch, are constantly deceived. Information received through fallible sources, from the nature of the case, must be open to suspicion.

But the mental powers and operations are themselves mutable and uncertain. Memory cannot be trusted implicitly. We find ourselves constantly mistaken in our recollection of things and events; and memory is absolutely necessary to any sort of knowledge. But even if it were never caught limping, how do we know that it is not always persistently and consistently false? What can we do but simply take what it tells us of the past as true without possibility of verification?

But in point of fact, all our ideas and opinions have in them undeniable sources of change and uncertainty. The mind itself is not the same from youth to age, and no one can tell at what stage it can best be trusted. Its grasp and flexibility undergo daily changes, through variation in health — through fits of passion, moods of despondency, the use of intoxicants and the effects of environment. All of these jangling voices cannot be true, and how can we be sure of any one of them? Then again, one's way of thinking is greatly modified by education, religion, birth, manners and customs, interest, and a thousand external circumstances. Some of these, or all of them must be distorting; and how shall we say that the Hindoo mother, the Howling Dervish, and the modern Thug, are not all equally justified in their conclusions?

The skeptic is not done yet. He asks—especially in the person of certain pseudo-scientists — why should not the mind, with all thought, be but an effect, like the blaze of a candle, or any other mechanical or chemical reaction? We can trace the motion from any external stimulus along the sensory and motor nerves to and from the nerve centres in the brain,—we can localize these centres and note their action: what more is needed? If the quiet, restful earth, and all that goes on in it is but the effect of mass and motion, why should we look further for an explanation of thought? Call it a phenomenon of matter and be content!

Or again,—and one can go no farther—even admitting that there is an infallible criterion of certitude, and that in spite of all that has been said, one could find it, how do we know that it would reveal that which is real — that the power behind nature and all phenomena has not so made us that we are compelled to see things quite otherwise than they are — that what we take for truth, and must take for truth is after all a delusion, unreal and false? The objects of the external world when seen through colored glasses, take the color of the glasses; why may not the whole thought-world appear what it is not, because it is seen through a medium which is forced upon us, and which gives a fictitious tone and character to truth; and so presents it to us as what it is not?

Even such a radical position as this is possible, and as an hypothesis it is irrefutable. We should have to be given another set of senses with which to examine this hyper-sensuous world, and compare the results with our present knowledge, before we could know whether we should see things as we do now, if we were angels or inhabitants of Mars or some other possible world. But even then we should only know that our new senses made things look to us as they would to an angel or a Martian; but the question would again obtrude itself — Does the angel or the Martian see things as they are? and so it would again and again, if we had a thousand sets of senses. No one set could do more than our present senses do, — that is, do just what is their business to do. The very highest created being — though he be only less than the All-Father, can know only as his powers and capacities enable him to know; and we are in no worse case. He would know more, doubtless, but he could not know with greater certainty. He too could doubt, — doubt as radically as man, — doubt being, as we have seen, an absolutely necessary condition of knowledge to any finite intelligence. If, then, we do not demand to be as Gods, — if we are to remain as created beings, — if there are to be created beings at all, there is no truth, — there can be no truth except as it is made known or revealed to the creature as he is, with and through the limitations which make him a creature. That is a false and absurd philosophy which attempts to carry one out of the scope of the actual and asks how things are or could be, to another order of beings than man. We are men, and in the man-world; and the very truth is the truth that man knows, or can know.

But it will be remembered that I said in the beginning of this discussion, that there is a practical limit to doubt unconsciously forced upon the unthinking at all times. By the necessity of man's nature, he cannot live in utter doubt and negation. After the sum of all actual doubt is reckoned up, there remains in his practical life a far vaster sum of unquestioned reality, through which, indeed, the very doubt and his own existence are made possible. Question and deny as we may, we have faith in our senses; and we show it every moment of our lives. There is no man who does not thoroughly believe that the earth is under his feet, and that if he does not take food he will die. There never was a man who did not know that there were other men about him, and that there were rights and duties growing out of the relations between them. There never was a man who, uncertain about a fact, did not know that if his means of investigation were sufficiently enlarged, the doubt would be removed. Granted that the senses do sometimes mislead us, does not this very knowledge emerge from the far deeper knowledge that they commonly do not? and are we not certain that if due precaution were taken, — if the obstructing or misleading elements were removed, the deception would go too? Is not the universal and necessary consciousness that we must not always trust the senses a certainty? and does it not carry with it the further and deeper conviction that there is an infallible criterion, if we can only be sufficiently informed? As Jouffroy, whose line of thought I have in good part followed, says, "The cause of our faculties deceiving us, is not the want of a *criterion* to distinguish the proper from the improper exercise of them, but carelessness or haste in not using this criterion."

With regard to the question of the accidents, or properties of bodies, and that 'thing-in-itself' which supports them, I have only to say here that the difficulty is one which presents itself only to the philosopher, and will come up for consideration further on. People at large are not troubled with any such abstraction, but innocently assume as the philosopher does also, when he is not philosophizing, that what lies spread out before him, as land and sea and sky, are simply what they seem. I also pass for the present that great domain of necessary truths, or presuppositions of our nature, as we shall be able to get at it better further on.

So much for the practical limits of skepticism. Let us now consider briefly the question from a purely logical standpoint. We are told that Descartes, the father of modern speculative thought, shut himself out from the world, and deliberately set himself to the work of carrying doubt to its direst limit. He found that there is one fact which stands out clear and distinct in the midst of one's most determined effort at denial, — and that is, the fact of one's own existence. Doubt as I may, I cannot doubt that I doubt. The ego is necessarily posited or affirmed in the very act of doubting. Consciousness is beforehand in forcing the knowledge of self upon me in the act of construing the notion of denial in any form, and in enunciating my conclusion the "self" is affirmed as a necessary condition: it is *I* that doubt.

The formula in which Descartes embodied this fundamental verity, 'I think, therefore I am' (*cogito ergo sum*) — has been much criticised and discussed. There is no reasonable question that the philosopher intended it not as an argument, but as an incontestable postulate. But it is a matter of no moment to us here what he may have intended; the truth remains. We need only the two words, 'I doubt.' They cover the whole range of skepticism; and it is logically impossible to entertain or formulate the expression of any sort of doubt whatever without positing incontestably the belief on the part of the doubter, in his own existence.

But this assertion, — 'I doubt,' — which lies at the threshold of all questionings, is pregnant with a further truth, equally important, and equally obvious. One could never have the slightest consciousness of self without the consciousness of the not-self. If the ego were in a state of absolute isolation — a unit, without that which is other than itself, even its own parts or limitations, there could be no possible variation in its modes or states, it could have no possible experience — there could be nothing to think about, and so no thought whatever, and no consciousness of existence. Thus the knowledge of the self carries with it necessarily, the knowledge of the non-self with its unending phenomena. And thus that flood of doubt, which we so freely admitted in the beginning, returns upon us in an overwhelming sense of certainty. It does not yet appear what the non-ego is; but that it is somewhat we cannot deny. We still know that the senses are not always to be relied upon, but we know also, that they are bearing their testimony, and it remains for the self to weigh it and determine. What we do know, and must know without question, is that we *think* there is an external world; and what one thinks and cannot by any possibility not think, one knows. Every sensation is at least a sensation, and in so far real: whether the external stimulus be indeed what we take it to be is quite another matter. It is just in this fact that flexibility is left us, and that we are saved from dead mechanism — a dire necessity without the possibility of thought or action. Of these two factors — the self and that which is not self — it will be seen, — and it cannot be logically disputed, — that the positive, living factor is the self. It would be premature to enter upon a discussion of the relation of these two factors at this stage of our investigation. It is enough to emphasize the fact that the world of mass and motion can be known only in thought; and that thought is the distinguishing characteristic of the self.

But now, what are we to understand by the *ego*, the *me*, the *self?* First, negatively (speaking for myself), I do not mean the body, nor the brain, nor any special organ of the body — I do not mean the memory, nor imagination, understanding, will, or consciousness, nor even what is commonly called mind or soul. I do mean all these — the whole self — all that goes to make up what we know as 'person' — in one sense compounded of parts, in another and higher sense, absolutely partless — a unit, not susceptible of any sort of fraction or division. We see in it a living exemplification of the problem about which philosophy, ancient and modern, has ever busied itself — the co-existence of the 'one and the many.' As 'many' it is composed of two chief factors — a marvellous mechanism, and an incomprehensible and dominant psychical energy: as 'one' it is a living and ineffable personality. The nature and existence of the mechanism, and of the psychic factor are known only through the personality which for each and every one of us is the one primordial and necessary fact of the universe.

CHAPTER II.

THE MECHANICAL BASIS OF PHENOMENA.

Modern science and the older learning. All science dominated by mechanics. Reducible to mathematical forms. Forestalled by Descartes. Hobbes. Leibnitz. Attitude of modern physicists. Metaphysical basis of science obvious to all thinkers.

IT is not surprising that men's minds should be in some ferment as to what is true, when one reflects upon the universal upturning in physical science, in the last two or three generations. Few facts, to say nothing of theories, are left unmolested. The old learning is so tattered and torn as to be no longer respectable. New discoveries and new hypotheses have crowded each other with such rapidity, that one feels fairly dazed when one thinks of it. Science has done such a mighty work already, and gives promise of so much to come, that it is not wonderful it has so fully engaged the attention of the age. It is not strange that so many hands seize the scalpel and microscope, — the battery and balances to push on the work, and that so many brave hearts put their trust in them as the only sure test of truth, — they are so definite and practical. It is quite natural that they who are once taken with the experimental method should think they have no time, and show so plainly that they have no patience, with the old hair-splitting, foggy metaphysic. And yet it will hardly do to cast contempt upon the old thinkers. The seductive path of positive science leads off into regions of speculative thought at numberless points ; and if Science does not already know that she is caught in the toils of metaphysics, it is only because she does not yet fully recognize her contact with the ultimate. Experimental science has been a trifle heady perhaps in taking leave of the old thinkers in the beginning; and she need not be surprised to find herself overtaken by them once more. Her revelations have rarely proved to be new, except in mere details, or new only in the sense that the multitude knows little of what the philosophers of the past have clearly seen, and definitely announced.

This could be abundantly verified by a study of even Greek philosophy, but that would carry us too far afield. We must content ourselves with a brief reference to the thinkers of modern times, but yet far enough in the past to have preceded by many years the scientific flood which seems to be sweeping everything before it.

The central principle of modern physics may be stated as follows: "All variations of matter, or all diversity of its forms, depend on motion": but these are not the words of an analytical physicist of our day, but of Descartes. He saw, as Professor Huxley says, that the discoveries of Galileo meant that the universe is governed by mechanical laws; while those of Harvey made it equally clear that the same laws preside over the operations of that portion of the world which is nearest to us, namely our own bodies. In his essay, "Traité de l'Homme," he arrives at "that purely mechanical view of vital phenomena towards which modern physiology is striving." Speaking of the mechanism of the circulation of the blood, he says, that the motion is as much the necessary result of the structure of the heart, as that of a clock is of the "force, the situation, and the figure, of its weight, and of its wheels." Nor does he stop with this. "The animal spirits" he says "resemble a very subtle fluid, or a very pure and vivid flame, and are continually generated in the heart, and ascend to the brain as a sort of reservoir. Hence they pass into the nerves, and are distributed to the muscles, causing contraction, or relaxation, according to the quantity."

He goes into details and describes the animal body as an automaton, — explaining the action of what we now call stimuli upon the sense-organs. He illustrates his meaning by likening the action to the mechanism of certain grottos and fountains in royal gardens. "The nerves of the machine" — he is speaking of an hypothetical human organism — " which I am describing may very well be compared to the pipes of these water-works; its muscles and its tendons to the other various engines and springs which seem to move them; its animal spirits to the water which impels them, of which the heart is the fountain; while the cavities of the brain are the central office. . . . The external objects which, by their mere presence, act upon the organs of the senses, and which by this means determine the corporal machine to move in many different ways, according as the parts of the brain are arranged, are like the strangers who, entering into some of the grottos of these water-works, unconsciously cause the movements which take place in their presence. For they cannot enter without treading upon certain planks so arranged that, for example, if they approach a bathing Diana, they cause her to hide among the reeds; and if they attempt to follow her, they see approaching a Neptune who threatens them with his trident; or if they try some other way, they cause some monster who vomits water into their faces, to dart out; or like contrivances, according to the fancy of the engineers who made them. And lastly, when the *rational soul* is lodged in the machine, it will have its principal seat in the brain, and will take the place of the engineer, who ought to be in that part of the works with which all the pipes are connected, when he wishes to increase, or to slacken, or in some way to alter their movements." He goes on further even than this, and includes in his mechanism the organs of " common sense and imagination " — indeed the most pronounced mechanical physiologist could go no further. Professor Huxley declares that the spirit of what he says "is exactly that of the most advanced physiology of the present day."

Thomas Hobbes, a very different thinker, clearly and in terms anticipated the results of the latest study in Physiological Psychology. He says:—"All the qualities called sensible are, in the object which causeth them, but so many motions of the matter by which it presseth on our organs diversely. Neither in us that are pressed are they anything else than divers motions; for motion produceth nothing but motion. . . . The cause of sense is the external body or object, which presseth the organ proper to each sense, either immediately, as in taste and touch, or mediately, as in hearing, seeing and smelling; which pressure, by the mediation of the nerves, and other strings and membranes of the body, continued inwards to the brain and heart, causeth there a resistance, or counter-pressure or endeavour . . . and because *going*, *speaking*, and the like voluntary motions, depend always upon a precedent thought of *whither*, *which way* and *what;* it is evident that the imagination [or idea] is the first internal beginning of all voluntary motion. And although unstudied men do not conceive any motion at all to be there, when the thing moved is invisible; or the space it is moved in is, for the shortness of it, insensible; yet that doth not hinder but that such motions are. These small beginnings of motion, within the body of man, before they appear in walking, speaking, striking, and other visible actions, are commonly called endeavour."

Professor Romanes, the distinguished English physicist, in commenting on this passage which he quotes in his Rede Lecture of '85, at the University of Cambridge, declares it to be in perfect accord with the best scientific thought of to-day — that it has now been proved beyond doubt to be only in virtue of the invisible movements which he inferred that the nervous system is enabled to perform its functions.

But while nothing could be more clear than the position of

Hobbes with regard to the mechanical constitution of the human body, he was neither original nor alone in his declarations. He was not only anticipated some two thousand years by Heraclitus and Empedocles and Democritus, and on down through Epicurus and Lucretius, but as we have seen Descartes had forestalled him by many years, and Leibnitz and Huygens, both of whom were quite as pronounced on the subject, were his contemporaries. Leibnitz says : " Everything in nature is effected mechanically " ; and he carried the doctrine of motion into all phases of his philosophy as a necessary postulate. Huygens, the father of the undulatory theory of light, declares that "all natural effects are, and must be, conceived mechanically, unless we are to renounce all hope of understanding anything in physics." A little later, but years before the descent of the present scientific avalanche, Father Boscovich put forth his theory of the construction of matter, in which motion and force alone give rise to all the phenomena of substance.

It was not, however, until after Priestly, Lavoisier, Dalton and the rest of them gave the world a new chemistry — not until after the announcement of the atomic theory, the establishment of the doctrine of the conservation of energy, and the mechanical equivalent of heat, that the latest phase of this venerable theory burst upon the thought of the world with its brilliant achievements. Kirchoff, Helmholtz, Clerk Maxwell, and a host of other mathematicians and physicists took up the inquiry, and seem to have settled the matter finally upon a mechanical basis as well in organic as in inorganic nature. The conclusions of Wundt may be taken as the accepted attitude in physiology, the most subtle domain of nature, and perhaps the ultimate reach of mechanics. He says : " The view that has now become dominant (in physiology), and is ordinarily designated as the mechanical or physical view, has its origin in the causal conception, long prevalent in the kindred departments of natural science, which regards nature as a single chain of causes and effects wherein the ultimate laws of causal action are the laws of mechanics. Physiology thus appears as a branch of applied physics, its problems being a reduction of vital phenomena to general physical laws, and thus ultimately to the fundamental laws of mechanics."

The statement of Du Bois Reymond is equally clear and positive — " Natural science more accurately expressed, scientific cognition of nature, or cognition of the natural world by the aid, and in the sense of theoretical physical science — is a reduction of the changes in the material world to motions of atoms caused by central forces independent of time, or as a resolution of the phenomena of nature into atomic mechanics. It is a fact of psychological experience, that, whenever such a reduction is successfully effected, our craving for causality is, for the time, wholly satisfied. The propositions of mechanics are reducible to mathematical forms, and carry with them the same apodictic certainty which belongs to the propositions of mathematics. When the changes in the material world have been reduced to a constant sum of potential and kinetic energy inherent in a constant mass of matter, there is nothing left in these changes for explanation."

There is no disposition on the part of any school of thought of fair respectability to question the general correctness of the statement embodied in the above quotations, nor to deny that the position of science on the general subject is, so far as it goes, a just explanation of the phenomena of nature; but we shall see, I hope, that there is a long way beyond the utmost reach of mass and motion, upon which it can take no step. It is not in the least surprising, however, that people who are unread in philosophy — and that means all except one here and there — should feel themselves in a state of spiritual asphyxia when they comprehend the sweep of the mechanical claims; and the case is made apparently far worse when the further researches of Darwin and Wallace, and that host of able collaborators in the evolutionary processes of nature are taken into account.

It does not seem necessary to our present purpose, at least at this point, to touch upon what is commonly known as the Darwinian theory; though I may remark in passing that a sound philosophy has no quarrel with it in its general aspects, nor as an explanation from the mechanical side of the phenomena about which it is concerned. We shall see, however, that it is seriously at fault in leaving out of sight or making little of the primordial factor of personality in its explanation of the processes of nature.

It is necessary that we should try to get a right notion of the latest results of the mechanical theory in its physiological aspect, not to combat it in any wise, but that we may be helped to right conclusions in the difficult inquiry as to the relations of Mechanism and Personality.

CHAPTER III.

PSYCHO-MECHANISMS.

The cell-theory modified. Protoplasmic movement. Max Schultze. Huxley. Uni-cellular organisms. Structural development. Professor Foster quoted. 'Metabolism.' Nervous system. Reflex action. Vivisection. Cerebral hemispheres. Effects of mutilations. Caution. Functions of different brain-areas not certainly determined.

THE cell-theory of physiology worked its way into favor gradually under the improved use of the microscope. From about 1835 it rapidly developed, until it culminated in the hands of, perhaps, Max Schultze in 1854.

This theory taught that all the parts of an animal or vegetable body, however different in appearance and structure, were built up of variously modified cells; and much emphasis was placed upon the cell-membrane, cell-contents and nucleus, with the idea that their modifications would account for divers organs, and their manifold functions.

This hope was doomed to disappointment, for within ten years of the researches of Schultze which had seemingly established the theory, that same histologist threw over the whole business of membranes as in anywise essential and retained only the enclosed living matter with its nucleus. He saw that the work done by the 'cell' was not so much the result of its cellular structure, as that that structure itself was a result of a formless living substance within, which he called protoplasm. The use of this word, protoplasm, now so common, dates from the publications of the researches of Schultze between 1861–63. Von Mohl had proposed the word in 1846, though without a proper conception of the thing. The credit of the discovery has to be divided among several. The French physicist Dujardin is no doubt entitled to the credit of priority in discovery, but the name he proposed—'sarcode'—did not adapt itself to the fancy, perhaps, and lacked the articulate roll of the other, and it has fallen into disuse.

The inquiry had been pushed back an important step by the recognition that, as Huxley puts it, the cells "are no more the producers of the vital phenomena than the shells scattered in orderly lines along the seabeach are the instruments by which the gravitative force of the moon acts upon the ocean. Like these, the cells mark only where the vital tides have been, and how they have acted." Hence arose, what may be called, the protoplasmic movement — "a movement which, throwing overboard altogether all conceptions of life as the outcome of organism, or the mechanical result of structural conditions, attempts to put physiology on the same footing as physics and chemistry, and regards all vital phenomena as the complex product of certain fundamental properties exhibited by matter, which, either from its intrinsic nature or from its existing in peculiar conditions, is known as living matter,—"mechanical contrivances in the form of organs serving only to modify in special ways the results of the exercise of these fundamental activities, and in no sense determining their initial development."

It is to be understood, then, that from the lowest possible forms of animal life in the protozoa, as well as in the lowest vegetable structures to the highest development in man, all tissues—all organs, nerves, muscles, cartilages—and whatever distinctions of tissue there may be, are built up out of the same protoplasmic unit.

It is not necessary to go into detail as to how structural forms are built up out of this structureless basis of life. In its simplest form, a single unit-mass constitutes a living being, as in the amœba. There cannot be said to be any distinction of parts; though perhaps this is not absolutely true: one such being seems to be elementary, presenting only one cellular development, but, while without organs or differentiations, it has nevertheless in itself many potencies, or capacities. All organs, looking backwards, find themselves despecialized, the unit-mass performing the functions of each as occasion requires. A unit-mass has the power of assimilation, by which it changes dead food into its living self: it has the power of movement or contractibility by which it adjusts itself to the performance of its functions as an animal, changing its form constantly, and it has sensitiveness or irritability which enables it to respond to external stimuli.

It is therefore very far from a unicum — oneness without marks or distinctions of form, states or power. It is simple indeed compared with the unending variety of specialized forms growing out of it, but as the meeting place, or common ground of all these it must, in its potentialities at least, be infinitely complex. It presents again the problem of 'the one and the many.'

I quote Professor Foster in *Encyclopedia Britannica* [Physiology], upon whom we may confidently rely: "The internal changes leading to these movements may begin, and the movements themselves be executed by any part of the uniform body, and they may take place without any obvious cause. So far from being always the mere passive results of the action of extrinsic forces, they may occur spontaneously, that is, without the coincidence of any recognizable disturbance whatever in the external conditions to which the body is exposed. They appear to be analogous to what in higher animals we speak of as acts of volition. They may, however, be provoked by changes in external conditions. A quiescent amœba may be excited to activity by the touch of some strange body, or by some other want, — what, in the ordinary language of Physiology, is spoken of as stimulus. The protoplasmic mass is not only mobile, but sensitive. When a stimulus is applied to one part of the surface a movement may commence in another and quite distant part of the body; that is to say, molecular disturbances appear to be propagated along its substance without visible change. The uniform protoplasmic mass of the amœba exhibits the rudiments of those attributes or powers which we described as being the fundamental characteristics of the muscular and nervous structures of the higher animals.

"These facts, and other considerations which might be brought forward, lead to the tentative conception of protoplasm as being a substance (if we may use the word in a somewhat loose sense) not only unstable in nature but subject to incessant change, existing indeed as the expression of incessant molecular, that is, chemical and physical, change, very much as a fountain is the expression of an incessant replacement of water. We may picture to ourselves this total change which we denote by the term 'metabolism' as consisting on the one hand of a downward series of changes (*katabolic* changes), a stair of many steps, in which more complex bodies are broken down with the setting free of energy into simpler and simpler waste bodies, and on the other hand of an upward series of changes (*anabolic* changes), also a stair of many steps, by which the dead food, of varying simplicity or complexity, is, with the further assumption of energy, built up into more and more complex bodies. The summit of this double stair we call 'protoplasm.' Whether we have the right to speak of it as a single body, in the chemical sense of the word, or as a mixture in some way of several bodies, whether we should regard it as the very summit of the double stair, or as embracing as well the topmost steps on either side, we cannot at present tell, — even if there be a single substance forming the summit, its existence is absolutely temporary: at one instant it is made, and at the next un-made. Matter which is passing through the phases of life rolls up the ascending steps to the top, and forthwith rolls down on the other side. . . . Further, the dead food, itself fairly but far from wholly stable in character, becomes more and more unstable as it rises into the more complex living material. It becomes more and more explosive, and when it reaches the summit its equilibrium is overthrown and it actually explodes. The whole downward stair of events seems in fact to be a series of explosions, by means of which the energy latent in the dead food, and augmented by the touches through which the dead food becomes living protoplasm, is set free. Some of this food energy is used up again within the material itself, in order to carry on this same vivification of dead food; the rest leaves the body as heat or motion. Sometimes the explosions are, so to speak, scattered, going off as it were irregularly throughout the material, like a quantity of gunpowder sprinkled over a surface, giving rise to innumerable minute puffs, but producing no massive visible effects. Sometimes they take place in unison, many occurring together, or in such rapid sequence that a summation of their effects is possible, as in gunpowder rammed into a charge, and we are then able to recognize their result as visible movement, or as appreciable rise of temperature."

The human body is composed of substances varying in molecular structure, each and all built up out of, or by means of protoplasm, and, however dissimilar in appearance and function, called tissues of the body. We are especially concerned with that most delicate system which may be called the borderland of the spirit world, and which certainly serves as the physical basis of all psychical action. This is called the nervous system, and without doubt it dominates the whole bodily organism. In immediate connection with it, and in necessary co-operation with it, we have the muscular system.

There are two general classes of nerves, the sensory and the motor (also called *afferent* and *efferent*). They are, so far as is known, of the same composition, being highly irritable fibres or strings, the sensory nerves normally conducting stimuli from the periphery or external surfaces to nerve-centres in the spinal cord, or the brain, and the motor system from the nerve-centres to the various muscles and other organs of the body. A stimulus being applied at the extremity or at any point along the course of a sensory nerve, an irritation or molecular movement, called a wave or impulse, is sent along the nerve till it reaches a nerve-centre, whence after a certain delay another impulse issues, out from the nerve-centre along the motor nerve, until it reaches, for example, the proper muscle, whereupon the muscle contracts. The action of the nerve extremity in response to an external stimulus, and the propagation of the impulse to the nerve-centre is due to the protoplasmic sensibility. The reaction is commonly looked upon as purely mechanical or non-mental in cases where the lower nerve-centres alone are concerned and is then called a reflex action. A large part of the bodily movement is accomplished thus without the intervention of thought or consciousness.

The well-known experiment with a frog is perhaps the best illustration of reflex action. If the hind foot of a decapitated frog is pinched, it withdraws the foot from the irritation, the sensory nerve transmitting the impulse to the nerve-centre in the spinal cord, and the motor nerve to the muscle, thus causing the reaction. If the irritation be made greater, the frog responds more strongly. This seems strange enough, but if the back of the frog be touched with an acid, it rubs it off with the foot on the same side. Cut off this foot and apply acid to the same spot — the frog makes an evident effort to use the stump of the amputated limb as before, but not succeeding, it makes use after some delay of the foot on the other side, and succeeds in rubbing it off, thus presenting the appearance of deliberative intelligent action.

The nerve-centres through which personal or psychical action is accomplished are found in the upper or cerebral portion of the brain, which in the higher vertebrate animals reaches a marked development, attaining in man an enormous proportional enlargement.

There are two cerebral hemispheres, connected by the great commissure or callosum, and so placed as to have their flat faces separated by a vertical partition, which passes from front to rear within the skull. They occupy the whole of the upper portion of the cranium down as far as the level of the eyebrows. Each hemisphere is composed of two substances quite different in appearance and structure,—the gray matter which is the upper and outer envelope, and the white matter which is the interior mass. The gray matter encroaches on the white irregularly, and is much corrugated and convoluted, the folds at many points extending far down into the white matter which supports it. While symmetrical in a general way, the convolutions and areas of the two hemispheres by no means correspond exactly. The gray matter is a congeries of nerve-cells, numbering, according to some estimates, more than a thousand millions—connected with each other and with the lower brain by a still larger number of nerve-fibres. The white matter which underlies the gray is composed essentially of these fibres; some of them connecting the cerebrum or fore-brain with the lower portions of the organ, some connecting the two hemispheres together, while others connect the different divisions of the same hemisphere. The gray matter is not found exclusively in the cortex of the hemispheres, but at all the levels between the latter and the termination of the spinal cord. Between, and occupying a central position with respect to the hemispheres, lie two paired masses of gray matter, called the optic thalami and the corpora striata, in which a large part of the nerve-fibres lose themselves, much as the wires of a telephone system converge at the central office. This system of the cerebral hemispheres is connected with the cerebellum, or hind-brain, and this with the medulla oblongata, which is an enlargement of the spinal cord, and so on down to the spinal cord itself. It is generally agreed that the reflex or automatic movements of the body are chiefly governed by this lower system, including the spinal cord itself. This cord is extremely complex, consisting of white and gray matter, but unlike the encephalic organs, the bulk of the white is exterior, entirely enclosing the gray matter within.

But there is no need — indeed it is out of the question to go into the details of physiology. The nervous system is too wonderfully complicated to be understood at a glance, and the foregoing sketch as well as what follows, must be taken as true, only as a mere outline. The simplicity of statement as to the sensory and motor nerves and the whole organism of the brain needs many qualifying statements in details, but they would in nowise affect our present purpose.

It is generally conceded that the mechanism which serves as the physical basis of conscious sensation is in the gray matter of the cerebral hemispheres, though the fact as to whether this is exclusively the case, is open to question. No one doubts that consciousness has a physical basis, and it is a matter of no great moment to the psychologist where, or what it is. It is not to be disputed that the cells of the gray matter of the cortex of the cerebrum are chiefly concerned in all well-defined psychical action, but it is difficult, perhaps impossible to find its limits. It seems rather, as has been held, that every nervous action affects in greater or less degree the entire system, and that there can be no line of separation drawn from finger tip to the superior frontal convolution.

The proofs that the nerve-cells in the gray matter of the cerebral hemispheres have chiefly to do with intelligent action are sufficiently clear. In the first place, it is found that, in the animal kingdom, the gradual development of the mass of this matter is fairly parallel with the corresponding rise in the scale of intelligence. The gray matter is found where the lowest well-marked volitional power discovers itself, and gains in volume and specialization as the power of conscious action in the ascending scale of animal life increases, reaching its culmination in man.

But the results of vivisection, carefully and cautiously practised upon the lower animals, and of the observations and experiments which opportunity has made possible in the human brain, all tend to establish the fact. As successive slices of the brain-matter are carefully removed from before backwards, an animal becomes more and more stupid, until at last all indications of perception and volition are gone. A pigeon so mutilated may live for months, but in a profound stupor without the slightest heed to what goes on about it. The animal still responds in a mechanical way to stimuli, and performs all the usual reflex adjustments in a sleepy fashion,— such as the use of the wings, when thrown into the air; sitting on its perch, and all its ordinary bodily movements; but it rarely moves unless stimulated from without.

Experiments have been made upon reptiles and mammals with like results.

The effects are quite different upon the removal of the cerebellum, or lower part of the brain. With a pigeon, says Professor McKendrick of Glasgow — "If the cerebellum be removed gradually by successive slices, there is a progressive effect upon locomotive action. On taking away the upper layer there is some weakness, and hesitation in gait. When the sections have reached the middle of the organ, the animal staggers much, and assists itself by its wings in walking. The sections being continued further, it is no longer able to preserve its equilibrium without the assistance of its wings and tail; its attempts to fly or walk resemble the fruitless efforts of a nestling, and the slightest touch knocks it over. At last, when the whole cerebellum is removed, it cannot support itself even with the aid of its wings and tail; it makes violent efforts to rise, but only rolls up and down; and then, fatigued with struggling, it remains for a few seconds at rest on its back or abdomen, and then again commences its vain struggles to rise and walk. Yet all the while sight and hearing are perfect. It attempts to escape, and appears to have all its sensations perfect. The results contrast very strongly with those of removing the cerebral lobes. . . . There is thus a loss of the power of co-ordination, or of regulation of movement, without the loss of sensibility, and hence it has been assumed that in some way or other the cerebellum acts as the co-ordinator of movements."

While the foregoing is undoubtedly true in the main, it must be accepted with some degree of caution. There is not a little conflicting evidence on the subject. Professor Ladd has brought the evidence together very fully in his work on Physiological Psychology. The specific functions of the cerebellum cannot be said to be yet fully determined. No disturbance of the sense of sight, hearing, or the muscular sense are certainly known to follow injuries to this organ when other parts of the brain remain uninjured, and cases are reported in which these senses were unimpaired in the total absence of the cerebellum. The case of a girl, Alexandrine Labrosse, is reported who was found at the autopsy to have had no cerebellum. In its place was a mere gelatinous membrane attached to the medulla by two peduncles of the same nature; yet she had no difficulty in co-ordinating her movements, and was in full possession of her senses. She fell easily, however, and had some difficulties of speech. Another case is reported of a man "whose entire cerebellum was changed into a brown purulent mass." He could walk, but in a tottering way. Another case, a woman dying at the age of sixty-nine, was found to have suffered an entire atrophy of all the gray substance of the cerebellum; yet she did not lose her muscular vigor, and could co-ordinate all her muscles, though her locomotion was disturbed and difficult.

CHAPTER IV.

PSYCHO-MECHANISMS (*continued*).

Professor Romanes quoted. Experiments of vivisection. Brain-localizations. Electrical stimulations. Rate of nerve-transmission. Time required for action of nerve-centres. Rate of nerve-vibrations. The sympathetic system. Functions. Independent of volition. Inhibition. Brain-development. Brain-mass. Different nationalities.

LET us return to the cerebral hemispheres. I make free use of the admirable summary given by Professor Romanes in his paper already referred to. If the gray matter of one hemisphere be removed, intelligent action is taken away from the corresponding (*i.e.* the opposite) side of the body, while it remains intact on the other side. For example, if a dog be deprived of one hemisphere, the eye which was supplied from it with nerve-fibres continues able to see, or to transmit impressions to the lower nerve-centre called the optic ganglion; for this eye will then mechanically follow the hand waved in front of it. But if the hand should hold a piece of meat, the dog will show no mental recognition of the meat, which of course it will immediately seize if exposed to the view of the other eye. The same thing is found to happen in the case of birds: on the injured side *sensation*, or the power of responding to a stimulus, remains intact; while perception, or the power of mental recognition, is destroyed.

This description applies to the gray matter of the cerebral hemispheres as a whole. But of course the question next arises whether it only acts as a whole, or whether there is any localization of different intellectual faculties in different parts of it.

Now in answer to this question, it has long been known that the faculty of speech is definitely localized in a part of the gray matter lying just behind the forehead; for when the part is injured, a man loses all power of expressing even the most simple ideas in spoken words, while the ideas themselves remain as clear as ever. It is remarkable that in each individual, only this part of one hemisphere appears to be used; and there is some evidence to show that left-handed persons use the opposite side from right-handed. Moreover, when the lesion occurs in the left hemisphere, either congenitally or during the first years of life, the right side apparently takes up the function.

Within the last few years the important discovery has been made, that by stimulating with electricity the surface of the gray matter of the hemispheres, muscular movements are evoked, and that certain patches of the gray matter, when thus stimulated, always throw into action the same groups of muscles. In other words, there are definite local areas of gray matter, which, when stimulated, throw into action definite groups of muscles. The available surface of the cerebral hemispheres has now been in large measure explored and mapped out with reference to these so-called motor-centres; and thus our knowledge of the neuro-muscular machinery of the higher animals (including man) has been very greatly furthered.

Here I [Professor Romanes] observe parenthetically that, as the brain is insentient to injuries inflicted upon its own substance, none of the experiments to which I have alluded entail any suffering to the animals experimented upon; and it is evident that the important information which has thus been gained could not have been gained by any other method. I may also observe that as these motor-centres occur in the gray matter of the hemispheres, a strong probability arises that they are not only the motor-centres, but also the volitional centres which originate the intellectual commands for the contraction of this and that group of muscles. Unfortunately we cannot interrogate an animal whether, when we stimulate a motor-centre, we arouse in the animal's mind an act of will to throw the corresponding group of muscles into action; but that those motor-centres are really centres of volition is pointed to by the fact that electrical stimuli have no longer any effect upon them when the mental faculties are suspended by anæsthetics, nor in the case of young animals when the mental faculties have not been sufficiently developed to admit of voluntary co-ordination among the muscles which are concerned. On the whole, then, it is not improbable that on stimulating artificially these motor-centres of the brain, a physiologist is actually playing from without, and at his own pleasure, upon the volition of the animals. The rate at which molecular movements travel through a nerve has been measured, and found to be about 100 feet per second, or somewhat more than a mile a minute, in the nerves of a frog. In the nerves of a mammal it is about twice as fast; so that if London were connected with New York by means of a mammalian nerve, instead of an electric cable, it would require nearly a whole day for the message to pass.

Next, the time has also been measured which is required by a nerve-centre to perform its part in a reflex action, when no thought or consciousness is involved. This time, in the case of the winking reflex, and apart from the time required for the passage of the molecular wave up and down the sensory and motor nerves, is about $\frac{1}{20}$ of a second. Such is the rate at which a nerve-centre conducts its operations when no consciousness or volition is involved. But when consciousness and volition are involved, or when the cerebral hemispheres are called into play, the time required is considerably greater. For the operations on the part of the hemispheres which are comprised in perceiving a simple sensation (such as an electrical shock) and the volitional act in signalling the perception, cannot be performed in less than about $\frac{1}{15}$ of a second, which is nearly twice as long as the time required by the lower nerve-centres for the performance of a reflex action. Other experiments prove that the more complex an act of perception, the more time is required for its performance. Thus when the experiment is made to consist not merely in signalling a perception, but in signalling one of two or more perceptions (such as an electric shock on one or other of the two hands; which of five letters is suddenly exposed to view, etc.), a longer time is required for the more complex process in determining which of the two or more expected stimuli is perceived, and which of the appropriate signals to make in response. The time consumed by the cerebral hemispheres in meeting a "dilemma" of this kind is from $\frac{1}{2}$ to $\frac{1}{20}$ of a second longer than that which they consume in the case of a simpler perception. Therefore, whenever mental operations are concerned, a relatively much greater time is required for a nerve-centre to perform its adjustments than when a merely mechanical or non-mental response is needed; and the more complex the mental operation, the more time is necessary. Such may be termed the physiology of deliberation.

So much, then, for the rate at which molecular movements travel through the nerves, and the times which the nerve-centres consume in performing their molecular adjustments. We may next consider the researches which have been very recently made upon the rates of these movements themselves, or the number of vibrations per second with which the particles of nervous matter oscillate.

If by means of a suitable apparatus, a muscle is made to record its own contractions, we find that during all the time it is in contraction, it is undergoing a vibratory movement at the rate of about nine pulsations per second. What is the meaning of this movement? The meaning is that the act of will in the brain, which serves as a stimulus to the contraction of the muscle, is accompanied by a vibratory movement in the gray matter of the brain; that this movement is going on at the rate of nine pulsations per second; and that the muscle is giving a separate or distinct contraction in response to every one of these nervous pulsations. That such is the true explanation of the rhythm in the muscle is proved by the fact, that if instead of contracting a muscle by an act of the will, it be contracted by means of a rapid series of electrical shocks playing upon its attached nerve, the record then furnished shows a similar trembling going on in the muscle as in the previous case; but the tremors of contraction are now no longer at the rate of nine per second: they correspond beat for beat with the interruption of the electrical current. That is to say, the muscle is responding separately to every separate stimulus which it receives through the nerve; and further experiment shows that it is able thus to keep time with the separate shocks, even though these be made to follow one another so rapidly as 1000 per second. Therefore we can have no doubt that the slow rhythm of nine per second under the influence of volitional stimulation, represents the rate at which the muscle is receiving so many separate impulses from the brain: the muscle is keeping time with the molecular vibrations going on in the cerebral hemispheres at the rate of nine beats per second. Careful tracings show that this rate cannot be increased by increasing the strength of the volitional stimulus; but some individuals — and those usually who are of quickest intelligence — display a somewhat quicker rate of rhythm, which may be as high as eleven per second. Moreover, it is found that by stimulating by strychnine any of the centres of reflex action, pretty nearly the same rate of rhythm is exhibited by the muscles thus thrown into contraction; so that all the nerve-cells in the body are thus shown to have in their vibrations pretty nearly the same period, and not to be able to vibrate with any other. For no matter how rapidly the electric shocks are allowed to play upon the gray matter of the cerebral hemispheres, as distinguished from the nerve-trunks proceeding from them to the muscles, the muscles always show the same rhythm of about nine beats per second : the nerve-cells, unlike the nerve-fibres, refuse to keep time with the electric shocks, and will only respond to them by vibrating at their own intrinsic rate of nine beats per second.

Thus much, then, for the rate of molecular vibration which goes on in the nerve-centres. But the rate of such vibration which goes on in sensory and motor nerves may be very much more rapid. For while a nerve-centre is only able to *originate* a vibration at the rate of about nine beats per second, a motor nerve, as we have already seen, is able to transmit a vibration of at least 1000 beats per second ; and a sensory nerve which at the surface of its expansion is able to respond differently to differences of musical pitch, of temperature, and even of color, is probably able to vibrate very much more rapidly even than this. We are not, indeed, entitled to conclude that the nerves of special sense vibrate in actual unison, or synchronize, with these external sources of stimulation ; but we are, it is thought, bound to conclude that they must vibrate in some numerical proportion to them (else we should not perceive objective differences in sound, temperature, or color) ; and even this implies that they are probably able to vibrate at some enormous rate. So far Professor Romanes.

The central nervous system is intimately connected with another set of ganglionic centres, and nerve-fibres distributed over the different parts of the body, but mutually connected with each other, called the sympathetic system. The principal centres of this system lie in the abdominal cavity near the spine, from which a series of trunks and branches known as the " solar plexus " radiate to the muscular walls of the intestinal canal, and to the various glandular organs connected with it. There are two other smaller plexuses, one in connection with the heart and great blood-vessels, and the other with the organs of reproduction and other viscera contained in the cavity of the pelvis, with smaller ramifications throughout the body. The action of this system is almost wholly automatic. It cannot be controlled by the will directly at all, but is easily affected indirectly by the emotions, particularly in relation to the heart and arteries. Everybody knows how easily the heart's movement is affected by the excitement of pleasure or apprehension, and how the blood mounts to the face in blushes or leaves it in pallor. All mental states react upon the sympathetic nerves, no doubt, so that digestion and health generally depend largely upon the cerebro-spinal system in its action upon the sympathetic. The functions of this system as a rule are performed silently and unperceived by consciousness. "The wheels of the inner life of the human machine move without noise," and only those who have studied the subject, or have been told, know anything of the far greater part of the functional action of the organs of the body.

In reflex action there is a distinction to be noted of great importance, as we shall see when we come to consider some of the phases of psychical phenomena. One class of such automatic action is wholly beyond the control of the will, and another set is directly under the dominion of volition. To the first class belong, as already said, all the activities of the sympathetic system, together with innumerable other mechanical functions which are absolutely necessary to life, and so are removed from all meddling on the part of the personality. To the second class belong many actions which, though in the main purely automatic, are yet susceptible of control. For example: if the bottom of one's foot be tickled, the reflex mechanism withdraws the foot from the irritation, but the will can set aside the automatic effort, and refuse to let the movement take place. This is an inhibition, and those elements of the personality whose functions are to overrule positive activities are called "inhibitory." These inhibitions probably depend upon mechanisms in the brain, since it is known that the removal of the brain is followed by greatly increased activity in the reflex centres of the spinal cord. Indeed, pure reflex action for the most part occurs only after the removal of the brain or in profound sleep. Any reflex movement which had its origin in the will can be inhibited, but those which cannot be incited by volition cannot be consciously overruled. There has been an effort made in these last years to discover a separate nerve-apparatus which is concerned in inhibitory action, but the existence of such a separate apparatus is not considered as established.

An interesting and important question arises as to the bearing of brain-development upon intellectual and moral power. It has been much discussed, and is still open to constant inquiry. There can be no doubt that the brain is the great unital mechanism of the human organism, and that its mass and specialization are fairly in the ratio of psychical power in the ascending scale of animal life. Its development has been traced in the utmost detail from the lowest orders in which the cerebral vesicles are discoverable up to the fully specialized and marvellously complex brain of man. Every step of the way — every division of what was single — every folding and tucking in, every differentiation, has been made an object of study, and it is not to be doubted that the structural brain-evolution is fairly parallel with the increase in psychical power of the beings whose nervous organisms are dominated by the encephalic organ. It would have been more amazing had it not proved to be so. As the central organ of a pure mechanism, it would have been out of all analogy and all meaning had not the mechanism of this organ increased in its specializations as the necessity of increased functional activity became necessary. As it is, we recognize in man the most marvellous mechanism to be found in the whole realm of nature, and in the brain the most marvellous contrivance of the Great Mechanician.

The weight of the brain-mass has no doubt something to do with the character of psychical powers, but this can only be affirmed in general terms. It is well known that many idiots and persons of low intellectual powers have had brain equal in weight to men of highest genius. A better mark, perhaps, of high intelligence is to be found in the number and delicacy of the convolutions of the cerebral hemispheres; but even this is still in the region of conjecture. It is probable, however, that the gray matter of the cerebral hemispheres is the centre of the mechanical element of thought, and yet connection of thought with the brain, or with the body in any form rests on faint sensations of effort, weariness and the like. That there is an immense amount of work done by the intercranial organs is witnessed to by the fact of the proportionally large amount of arterial blood necessary to enable them to fulfil their functions. The entire encephalic organ weighs only about $\frac{1}{45}$ of the body, and yet the amount of blood which it can hold is said to be about $\frac{1}{8}$ of the whole.

Comparative anatomy has been busy with the problem as to the relative amount of brain-mass in different nationalities, but so far the results have not proved conclusive, though the opinion is generally held that the average weight in the civilized races is somewhat greater than that in the savage races. The average weight of the brain of an adult male European is about 49 ounces, while it is said that the average of the savages in Oceanica and Africa falls somewhat below. The brain of women is about ten per cent less in weight than that of men. A portion of this is doubtless due to the differences in the height and weight of the body, and he would be a bold man who, in these days, would dare to hint that the margin left is not quite made up by other unreckoned factors. A large amount of the brain-substance may be lost without affecting, so far as can be observed, the intellectual character of the patient.

The brain attains nearly its full size quite early in life — by the age of eight according to accepted authorities, but it is thought by later investigators that this period is too short, and should be extended to the age of fifteen. It continues, however, to increase in weight till 30 or 35 years, or possibly even longer. After 60 it begins to diminish in weight. The weights of the brains of a number of distinguished men are on record: the brain of Cuvier weighed 65.4 oz.; Dr. Abercrombie's, 63 oz.; Spurzheim's, 55 oz.; Louis Agassiz's, 52.7 oz.; Dr. Chalmers', 53 oz.; Webster's, 53.5 oz.; Thackeray's, 58.5 oz.; Byron's, 63 oz. These are all much above the average; but many men of great mental powers have had brains not at all remarkable for weight or size; while the brain weights of many idiots have run fairly up to the maximum.

CHAPTER V.

THE SENSES — TOUCH, TASTE, AND SMELL.

The specific senses. Number indefinite. Touch fundamental. Pressure. End-organs. Threshold value. Weber's and Fechner's law. 'Local signs.' Pressure spots. Temperature spots. End-organs of taste. Stimuli. Classification. Sense of smell. End-organs of smell. Stimuli. Classification. Muscular sense. End-organs of motion.

LET us proceed now to look into the marvellous mechanism by which the personality is brought in contact with the external world. First, we have what are called the 'special senses' — that is, the old-time five senses, touch, taste, smell, hearing and sight; but there are vigorous claims on the part of physiology for the recognition of a number of other senses as specific factors in sense-perception; — notably what is called the 'muscular sense' — the 'temperature sense' — the 'general sense' (*sensus-communis*) — the sense of 'pain and pleasure,' and the sense of 'innervation and weariness.' Whether all these, or any of them, are entitled to rank as distinct senses with separate organs, is yet in question. Those investigators who hold the negative contend that they are but combinations of simple elements or states of consciousness. Leaving this question aside let us take up the recognized special senses separately, and we shall find as we go where the disputed points lie, and what is claimed.

Since the days of Democritus, Touch has been held to be in some sort the fundamental sense. In the light of modern research this is but confirmed and explained. All the senses, as we have seen, are immediately dependent upon motion for their mechanical action; and since the bodily organs must have motion communicated to them by stimuli of one sort or another, contact is a fundamental necessity in the establishment of reactions in the body. But this general sense of mere contact is far too wide in its scope to be called touch proper. In point of fact, there is no consciousness of contact in any of the senses, except in the specific sense which has this for its characteristic; that is to say, specific Touch is that sense which has for its note or mark the consciousness of actual contact from pressure or impact.

What is the physiological solution of this sense of pressure? Is simple contact of any sort, and with any part of the body all that is necessary to rouse this sense of touch? By no means. Some parts of the body, notably the brain, are wholly insentient. Even contact with a sensory nerve along the different points of its course, usually gives rise to pain which quite obliterates any specific sense of touch. Pressure sensations are normally due to the excitation of the end-organs of the sensory nerves, which are found in the skin, but distributed by no means uniformly in it.

Histological research is still busy with the general problem of the nature and functions of these end-organs. Two classes of them are, however, clearly recognized in which the sensory nerves terminate in the skin, one called 'end-fibres,' and the other 'end-bulbs.' We cannot enter upon the refinements and disputes of the many investigators. It is enough for our purpose to know that the 'end-bulbs' or 'corpuscles' of Pacini — the first discovered — are minute plum-shaped enlargements at the ends of medullated nerves. They are especially in the palms of the hands and fingers, in the soles of the feet, the toes, and to some extent in the neck, arms and other parts of the body. They can sometimes be seen by the naked eye, and are from $\frac{1}{10}$ to $\frac{1}{20}$ of an inch long.

There are a number of other classes of end-bulbs besides

these of Pacini far more remote, and of various structures. The nerve-fibrils are still more minute and more numerous, found in coils, spread out like rootlets, or with hair-like non-medullated endings. All these different classes are inter-mixed, much more thickly clustered in certain places so that the whole skin is thus made the organ of tactual sensibility. The exact functions of these several classes of nerve-endings are not yet certainly determined.

Contact sensations are divided into what is called Passive Touch and Active Touch. By Passive Touch is meant the sensation due to pressure applied to the body with the least possible motion of either the part touched or the object touching. Now it is found that not all pressure so applied is discoverable in consciousness. The pressure must reach a certain intensity, varying for different parts of the body, before it is discovered by the person touched. The least degree of pressure which can be felt is called the 'threshold value'; as though consciousness were an open door, and sensation had to rise to a certain height before it could flow in.

Investigators tell us that the least weight which can be felt on the forehead, temples and back of the hand is 0.002 gramme; in the nose, lips, chin, etc., 0.005; in the skin of the heel, 1 gramme.

But it is found that not every change in the degree of pressure is discoverable in sensation. Sensations, even of the same kind, do not always shade into each other. They appear to go *per saltum*. For example, if a weight of three ounces be placed upon a sensitive portion of the hand, or applied as a pressure, and it be increased to $3\frac{1}{100}$, the patient cannot discover that there has been any change. It may be increased to $3\frac{1}{75}$, to $3\frac{1}{50}$, and still sensation will not so change as to be discoverable in consciousness; but when it reaches about $3\frac{1}{40}$ the difference is felt at once. This ratio 3 to $3\frac{1}{40}$ is approximately constant for passive touch with this weight, but varies considerably when much larger or smaller weights are used. This difference which is necessary to produce a change in sensation is called the 'difference threshold,' and for passive touch, is about as 3 to $3\frac{1}{40}$ under the conditions given above.

It is thought by physiologists that something like this relation runs through all the senses, the ratio being different for the several senses and varying according to the intensity of the stimuli. There are difficulties in the way of determining what these several ratios are in some of the senses, from certain fluctuations not accounted for; but there is sufficient ground to think that the main fact nevertheless obtains. The principle is known as 'Weber's Law' and may be enunciated thus: the difference between any two stimuli must attain a constant ratio to produce successive equal steps in sensation — *i.e.* if sensation increases in an arithmetical ratio, the stimuli must increase in a geometrical ratio. This is also known in this modified form as Fechner's Law; who, however, states it in a more precise mathematical way: the intensity of sensation varies with the logarithm of the stimulus.

But not only do we distinguish sensations of touch as differing in intensity, but also as having place in the body. This fact of localizing stimuli is also quite marked in sight, and in less degree in hearing. There is considerable difference of opinion as to how this comes about, but the fact is not disputed.

It is plain that if a sensation, as of 'red' were confined to the nerve-sense of 'redness' there could be no objective meaning in it; but we not only have the sensation of color, but of color localized somewhere. So in touch: if we simply felt pressure in a general way, we should not know whether it were hand or foot which is pressed. This has nothing primarily to do with the question as to how we learn to localize sensations; but simply with the fundamental differences in sensation through which it becomes possible even to learn place-quality in sensation.

The theory which meets most favor is that of 'local signs'

proposed by Lotze. It is briefly, that with the sensation proper, there comes to consciousness through the organ of sense, a somewhat which serves to distinguish the sensation as due to a stimulus emanating from a definite place in space, and which he calls its 'local sign.' What the mechanical basis of these local signs may be is still open to question.

The sense of locality in the skin varies greatly in different areas. The discriminative sensibility in the skin has been carefully studied, the first work in this direction having been done by E. H. Weber. He used a pair of dividers, the blunted points of which were brought in contact with the skin, and the least distance apart of these points, which produced two distinct sensations, was taken as the measure of sensitiveness of the area. The tip of the tongue is found to be most sensitive, the two points being distinguishable as distinct when only about 0.04 of an inch apart. The points of the fingers come next, the distance being about twice as great; the inner or red part of the lips, a fifth of an inch; and so increasing in distance until in the middle of the back, the upper arm and leg, the distance has to be about two inches and a half.

Later investigators have established the fact that there are great individual differences in this discrimination of two points, some persons not being more than one-fourth as sensitive as others, though the relative acuteness for different parts of the body remains substantially the same. It is also found that the delicacy of discrimination is susceptible of very considerable, and even rapid cultivation, especially in certain areas of the body, notably the fingers.

These conclusions of Weber and his immediate followers have been greatly modified by the later studies of physiologists, especially Goldscheider. They discover what are called 'pressure spots.' Between these it is impossible to excite the sensation of pressure at all. There is sensation indeed, but dull, indefinable and expressionless.

The discriminating sensibility of touch is greatly augmented by movement, or successive changes. The threshold value and the difference threshold are much more refined, but the same constancy of ratios is maintained. It is thought that the experiments by Weber, while revealing what is quite true with regard to the skin, and its reaction practically, were not conducted with sufficient nicety to distinguish between a number of factors involved in his experiments.

It will be as well to mention here, while speaking of the skin and its functions, the remarkable discoveries with regard to the sensation of temperature. It may be now considered as established that only certain definite points of the skin are sensitive to heat and cold; and what is more astonishing, these points seem to be distinct; that is to say, the 'heat-spots' are not sensitive to cold, nor are the 'cold-spots' to heat. What is more, these spots are insensible to pain — even the pain which results from heat and cold. They may be pierced by a needle without sensation. In the words of Professor Ladd: "By using a machine which locates the stimulus microscopically, the topography of the skin may be mapped out, and extremely minute spots indicated which respond to irritation with sensations of pain, of pressure, of cold and heat, respectively.

"These different kinds of sensation spots appear never to be superposed; nor are they located alike on the symmetrical parts of the same individual, or on the corresponding parts of different individuals. . . .

"Heat-spots are on the whole less abundant than cold-spots; but in parts of the body where the skin is most sensitive to either heat or cold, the corresponding class of spots is relatively frequent. Temperature spots may be divided into first-class and second-class (so Goldscheider) — according to the strength with which they react on moderate stimulation. Some spots are roused only by excessive temperatures. The same object feels cool to one spot, ice-cold to another."

While not yet fully established, the reasons for considering temperature to be a specific sense with its own special nerves and end-organs are very strong. Just what these end-organs are is not yet certain. They are quite distinct from those which give rise to pain, since they may be in full possession of their sensibility, while those which give rise to pain in the same area are rendered insentient. Cocaine, which renders an area to which it is applied insensible to pain, leaves the sensation of temperature unaffected.

The end-organs of taste are placed at the entrance of the alimentary canal, and are probably confined to the upper surface of the back part of the tongue, the edges and the tip of the tongue, and the front part of the soft palate, though some physiologists claim that there are other areas susceptible to taste stimulation. The middle of the tongue and the surface of the hard palate are insensible to taste stimuli. What this stimulus is, nobody knows. Histologists, however, are substantially agreed as to the structure and disposition of the end-organs themselves. They are found in the papillæ of the taste areas, and are flask-shaped with a short neck which is towards the outer surface. They are very minute, imbedded in the mucous membrane, surrounded by epithelial cells, the opening or pore being from twelve-hundredths to four-thousandths of an inch in diameter. They are called 'gustatory-flasks' or 'taste-bulbs' or 'knobs' and are quite complex, each bulb being composed of from fifteen to thirty long, slender cells curving in at the top like the petals of a bud.

Only liquid bodies, or such as are in some measure soluble, act as stimuli to these end-organs, and not even all such bodies have taste. As to whether gases have taste or not may be considered an open question inclining to the negative.

A number of investigators hold that the sense can be excited by mechanical means; as for example, that pressure on the back part of the tongue will produce a bitter, and tapping gently and repeatedly on the tip, a saltish sensation. This latter effect can be easily verified and seems to be as claimed. Heat is not a stimulus, but after much debate it is now established that electricity is. If the cathode is placed upon the upper surface of the end of the tongue a sensation described as sourish-metallic, bitterish-metallic, etc., is said to be produced, while the anode in the same spot produces a somewhat similar but distinguishable taste. Rosenthal, as Professor Ladd states, finds that "when a chain of four persons is arranged in such manner as to send a current of electricity through the tongue of one, the eyeball of another, and the muscles of a frog-preparation held by two of the four, the same current will cause simultaneously an acid taste, a flash of light, and a movement of the animal's muscles."

It is impossible to make any scientific classification of the different kinds of tastes. The general rough classification is into sweet, sour, salt, and bitter. The sense of smell enters so largely into many shades of taste, that it is often difficult to say offhand to which sense a dominant quality is due. The muscular sense is also often involved. Intense stimulation of the taste-organs excites marked effects in the muscular system. The sympathetic nerves also are often involved, as for example, in nausea. Pungent, alkaline, astringent, and metallic tastes are held to be combinations, and generally flavors of bodies are complex. The sense varies through a considerable range in different persons, and is susceptible of a high degree of cultivation. Astonishing stories are told of the power of professional tastes, as of tea, liquors, etc., to discriminate nice shades of excellence and detect foreign substances.

The sense of smell is physiologically the least known of all the senses. The end-organs are found in the mucous membrane of the upper region of the nasal cavity and guard the opening of the respiratory tract. The olfactory region (*regio olfactoria*) has a thicker mucous membrane than the lower region (*regio respiratoria*) of the cavity. It is of a yellow or brownish-red character and in it are the olfactory cells. The cells are long, narrow, and spindle-shaped.

Only substances in a gaseous or volatilized state can stimulate the organs of smell, and even then the conditions must be favorable. The nostrils may be filled with any odorous particles, as eau de Cologne or sulphuretted hydrogen, without any sense of odor if there is no movement of the particles, and even then the movement must be inspiratory, or from without into the lungs. In the contrary movement of the gas or vapor, that is, in the expiratory movements of the lungs, there is no action on the part of these organs.

Thermal excitations do not give rise to smell, and the current opinion is that the sense cannot be produced by electrical or mechanical means, but the matter is by no means settled.

The degree of fineness of odoriferous particles is wonderful. If the air bearing an odor be filtered through a tube filled with cotton wool, and inserted in the nose, the smell is still discoverable. It is said that by this means organisms which cause putrefaction and fermentation of $\frac{1}{100,000}$ of an inch in diameter are removed. A grain of musk will scent a room for years, and yet at the expiration of that time no discoverable diminution of weight can be detected. One part of sulphuretted hydrogen in a million parts of air can be distinguished.

There is every reason to believe that the lower animals have a power of smell far beyond that possessed by man in his normal state. The dog and cat are especially furnished with information through this sense. There is a case on record of a boy — James Mitchell — born blind, deaf, and dumb, who could at once distinguish a stranger by this sense, and recognized his acquaintances by their distinguishing odors, and as we shall see, hypnotic patients seem to possess a power of smell which is marvellous.

The classifications of odors have no scientific basis. The

same substance gives rise to different odors, to different persons, or, if the odors remain the same, their agreeable, or offensive character differs widely, not only in different people but with the same person at different times. Certain effects, commonly accounted smells, do not properly belong to this sense; such as those called pungent, sharp, and irritating. It is claimed by some that even an acid has no smell proper, but that its action is due to mechanic irritation.

The first effect of smell is strongest. The first sniff of a rose is sweetest and most intense; after being inhaled for a moment the scent appears to die away. This may be due, it has been suggested, to a rapid coating of the olfactory membrane, but is more probably a subjective effect. It is thought that 'odors of animal effluvia are of a higher specific gravity than the air, and do not readily diffuse, — a fact which may account for the pointer and bloodhound keeping their noses to the ground.'

CHAPTER VI.

THE SENSE OF HEARING.

The ear. Structure. Corti's organ. Theories. Physical basis of sound. Intensity, pitch, quality. Illustrations. Partials. Tyndall quoted. Difference in people's sensibility. Powers of discrimination. Range. The human voice.

THE organ of hearing, the ear, is an extremely complex mechanism. It consists of an external, a middle and an inner ear. The functions of the external ear, — consisting of the auricle or convoluted cartilage at the side of the head, and the crooked tubular passage (*external meatus*) do not seem to be important except for admitting vibrations of the air to the mechanism within. The external cavity may be entirely filled with wax or tallow, and, if a passage by a tube be left to the middle ear, sounds are rather more distinctly heard; though it is thought that certain minor modifications of tones are affected through the external apparatus. The external meatus is a perfect protection to the ear drum, being a passage one and a quarter inches long, and somewhat bent downwards and backwards.

The *tympanum*, or drum or middle ear, is a chamber irregular in form, across which is stretched the tympanic membrane — itself complex, consisting of three distinct layers. Immediately behind this membrane are three small, curiously shaped bones called the hammer, the anvil and the stirrup stretching across the cavity from the tympanic membrane to the inner wall. The handle of the hammer (*malleus*) is connected with the middle of the tympanic membrane, and its head fits into a cavity of the anvil (*incus*) and has a delicate articulatory movement with it. One of the two processes of the incus ends in a rounded head and articulates with the *stapes* or stirrup. At its interior wall the middle ear opens into the *Eustachian tube*, a canal communicating with the nasal compartment of the pharynx. The office of the middle ear seems to be chiefly to transmit vibrations to the inner ear, though doubtless it performs important modifying functions not yet fully understood.

The internal ear, called also the *labyrinth* from its complexity, is the part of the auditory apparatus in which the true end-organs of hearing are placed. Without attempting to enter upon details it must suffice to say that it consists of three parts known as the vestibule, the semicircular canals, and the cochlea. Of these the cochlea is by far the most complex. The osseous cochlea is a tube wound two and three-quarters times round a pillar as an axis, like the shell of a snail, both pillar and tube diminishing rapidly in diameter from base to apex. The enclosed membranous mechanism here is marvellous in the extreme; but we pass on to the remarkable arrangement of cells discovered by the Marchese Corti, and so called the organ of *Corti*. It is a membrane composed in part of fibres which are stretched at right angles to the longer axis, *i.e.* radially. This membrane is furnished with 10,500 'rods' or 'pillars of Corti,' arranged in an inner and outer row and increasing in length from the base to the apex of the cochlea. Each set of rods has a row of hair-cells so-called nearly parallel with it, and these are covered with a delicate perforated membrane. The hair-cells communicate with the auditory nerve and so with the brain.

There is considerable obscurity yet, as to the exact action of this auditory apparatus, but there is little doubt that it is purely mechanical. At one time Helmholtz thought that the rods of Corti responded to the vibrations communicated to them, such vibrations throwing into action those fibres which were in sympathetic accord; and since the estimated number of these would allow thirty-three filaments for each semitone through the whole range of audition, there would be nearly enough to answer to every shade of tone; but even if the number fell short, it was shown that nicer shades were possible by the composite action of the two filaments between which the sound might fall. Recent histological researches, however, have led to some modifications of this theory, especially as it is found that there are no rods of Corti in the cochlea of birds, and it can hardly be doubted that they have an appreciation of pitch. "Hensen and Helmholtz have now suggested the view that not only may the segments of the 'membrana basilaris' be stretched more in the radial than in the longitudinal direction, but different segments may be stretched radially with different degrees of tension, so as to resemble a series of tense strings of gradually increasing length. Each string would then resound to a vibration of a particular pitch communicated to it. The exact mechanism of the hair-cells, and of the *membrana reticularis*, which looks like a damping apparatus, is unknown."

Sounds may be divided into two classes, noises and musical tones. Whether there are separate end-organs through which these two classes are conveyed to the brain-cells is not definitely known; but, however this may be, it is certain that the stimuli in either class is motion which causes the air to be thrown into agitation, or tremor, as by a blow or oscillation of some external body. The fact that there is no discoverable relation between the power of distinguishing noises and of appreciating musical tones, gives plausibility to the theory that they are due to distinct mechanisms in the nerve-endings. The two classes are clearly distinguished mechanically at all events. In tones there is found to be regular recurrent motion, the period or time of the recurrence being uniform, and so called periodic. In noises there is an absence of this element, but instead, confusion and lack of uniformity. Noises, however, can be detected in almost all musical tones, as the scraping of the bow in the violin, the whir of the air in the flute, and the rattle of action and strings in the piano. So also in noises, tones can almost always be detected by a trained ear, as in the ring of a hammer, the creaking wheels, and the resonance of an explosion.

Any regular recurrent motion gives rise to a tone. A sufficient illustration, though not the most perfect, is found in Savart's machine, which is simply a toothed wheel, the teeth striking upon a bit of card-board in the revolution of the wheel. When the wheel moves slowly so that the taps reach, say, 40 or 50 in a second, a very low tone is produced, and as the velocity is increased the tone runs up till it reaches the highest pitch. This machine presents the mechanical action most perfectly to the eye, but the 'Siren,' which simply makes and breaks a current of air, is far more perfect.

The Science of Acoustics has for its object the development of the mechanical phenomena of periodic motion in sound, and the mathematicians have given a very complete analysis of the whole subject. We are compelled to content ourselves with a brief outline.

A body under a central or directed force must move according to the laws of mechanics, in one or other of the class of curves known as conics, so called because they may all be cut from the surface of a cone by a plane. Gravitation is such a force, and the motions of the bodies of the solar system are examples of periodic motion. A common pendulum is the simplest exemplification of such motion. Galileo discovered from a swinging lamp in a church, it is said, that, although the arc through which the pendulum swings gradually loses in length on account of the resistance of the atmosphere and friction, the beat or time of the swing remains constant. So in all periodic motion, the time in which an excursion is made is entirely independent of the amplitude, or extent, of the path. This principle lies at the bottom of all sustained or musical tones.

Now elasticity, or the energy of restitution when a particle is disturbed in an elastic medium, is a directed force, and for small distances this law of periodic motion holds for all oscillations. A displaced particle is driven back by this force of restitution, but by virtue of the kinetic energy generated in its return, it passes its original place of rest, and swings to the other side, and so back and forth, until the moving energy is exhausted, but taking just as long a time to make its final and least movement as it did for the first and greatest. This swing or pendulous motion (called vibration) in conic curves (a straight line is a particular case of an ellipse) is the mechanical basis of the undulatory or wave theory of sound, light, heat, and, we may say, of all physical phenomena.

To affect the ear, the air, which is an elastic medium, must be thrown into a state of agitation, and the impact of the air particles upon the tympanic membrane causes it to vibrate, and these vibrations are transmitted through the wonderful mechanisms of the ear until they finally reach the end-organs of the auditory nerve, and are thence conveyed by nerve-fibres to the brain.

Now the elementary motions, or excursions of the molecules which, transmitted, give rise to a pulse or wave, are performed in a very small space. These minute motions are characterized by three distinct variations with respect to the path or orbit over which they move. First, the orbit may be large or small. If large, the velocity, and consequently the moving energy, must be great compared with a smaller path and slower motion, the time being constant. This causes the impact or blow on the tympanic membrane to be greater for large orbits than for small: that is, the loudness or intensity of sound depends upon the amplitude of the impinging air particles.

The second variation is in time. The orbital or periodic times of the moving molecules may be greater or less according to the rapidity of movement of the body which gives rise to the vibrations. The greater the number of vibrations in any given time, as a second, the higher or sharper the tone. This is what is called 'pitch.' The lower limit, or gravest sound audible to most ears, is about 30 vibrations to the second: the upper limit, or most acute sound, has about 30,000 to the second.

The third variation is in the form of the path in which the molecules move. If a pendulum formed of a cord and weight attached be started to vibrate, the time of vibration will remain constant so long as the length of the cord is unchanged, though the pendulum bob be made to describe any sort of figure (as it would be seen from above), as for example a straight line, an ellipse, or a figure eight. There may be any number of small motions superposed upon the main path, like the motions of the moon, which, while moving around the sun, has at the same time, a motion around the earth, as well as divers other minute perturbations. This variation in the form of the path gives rise to what is called 'character,' 'timbre,' 'klang,' or quality of tone.

To illustrate: if a tuning-fork on its stand be struck, the rapidity of movement of the two prongs will always be the same for the same fork, whether the blow be soft or hard. This gives the same time for excursions of the air particles, and so the 'pitch' and the 'timbre' remain constant for that particular fork. Hence it is that tuning-forks have to be selected for the particular pitch, or note, they are desired to give. But if the same fork be struck softly, and then with greater force, there is a marked difference in the intensity of the sound. The swing of the prongs is greater, and the excursions of the air particles are of greater amplitude, so that the effect on the ear is to give greater loudness for the heavy stroke than for the soft one; that is, the 'intensity' differs for the two notes.

Again, if two different tuning-forks are making, one 528 vibrations in a second, and the other 792, the first will be middle C and the second G above, and almost any ear will discover that the first is more grave than the other. This is a difference in 'pitch.'

Once more, if a particular note, as middle C, be sounded on two different kinds of instruments, as a flute and a clarionet, or a violin and a piano, although they may be exactly the same pitch and loudness, they are easily distinguishable. This is called 'timbre' or 'quality.' It is due to the difference in the form of the path in which the excursion of the air particles is performed; and this difference is caused by superposed movements upon the elementary path, called 'overtones' or 'partials': that is to say, along with the principle tone there are other tones sounding at the same time, and it is hardly possible to get a tone perfectly simple or free from these riders; which are also called 'harmonics.' When a note is most nearly simple it is thin, meagre, and insipid. The lower harmonics give to the fundamental tone richness and fulness, while the higher give brilliancy and thrill.

The manner in which bodies break up into multiform vibrating segments cannot be better illustrated, perhaps, than by the following extract from one of Dr. Tyndall's lectures on sound. He says: "We are now prepared to appreciate an extremely beautiful experiment, for which we are indebted to Professor Wheatstone, and which I am now able to make before you. In a room underneath this, and separated from it by two floors, is a piano. Through the two floors passes a tin tube $2\frac{1}{2}$ inches in diameter, and along the axis of this tube passes a rod of deal, the end of which emerges from the floor in front of the lecture table. The rod is clasped by india-rubber bands, which entirely close the tin tube. The lower end of the rod rests upon the sound-board of the piano, its upper end being exposed before you. An artist is at this moment engaged at the instrument, but you hear no sound. I place this violin upon the end of the rod; the violin becomes instantly musical, not, however, with the vibrations of its own strings but with those of the piano. I remove the violin, the sound ceases; I put in its place a guitar, and the music revives. For the violin and guitar I substitute this plain wooden tray; it is also rendered musical. Here, finally, is a harp, against the sound-board of which I cause the end of the deal rod to press; every note of the piano is reproduced before you. I lift the harp so as to break its connection with the piano, the sound vanishes; but the moment I cause the sound-board to press upon the rod, the music is restored. The sound of the piano so far resembles that of the harp that it is hard to resist the impression that the music you hear is that of the latter instrument. An uneducated person might well believe that witchcraft is concerned in the production of this music.

"What a curious transferrence of action is here presented to the mind! At the command of the musician's will his fingers strike the keys; the hammers strike the strings, by which the rude mechanical shock is shivered into tremors. The vibrations are communicated to the sound-board of the piano. Upon that board rests the end of the deal rod, thinned off to a sharp edge to make it fit more easily between the wires. Through this edge, and afterwards along the rod, are poured with unfailing precision the entangled pulsations produced by the shocks of those ten agile fingers. To the sound-board of the harp before you the rod faithfully delivers up the vibrations of which it is the vehicle. This sound-board transfers the motion to the air, curving it and chasing it into forms so transcendently complicated that confusion alone could be anticipated from the shocks and jostle of the sonorous waves. But the marvellous human ear accepts every feature of the motion; and all the strife and struggle and confusion melt finally into music upon the brain."

There is a marked difference in the sensibility of different people to sounds, although they may have what is called perfect hearing. This is especially true in the higher range. When the test is made, certain people are surprised to find that their ears fail to respond to tones which are distinctly heard by others not supposed to be any better. What the range is in the lower animal kingdom cannot be satisfactorily ascertained, but there seems to be good reason for supposing that there may be a whole world of sound, especially for insects, which we count silence.

The niceness of discrimination in the differences of tone in a practised ear is amazing. Some musicians can detect about the $\frac{1}{5}$ of a vibration; and on the other hand, it is not uncommon to find people who cannot distinguish semitones, or even tones. When a note has a number of vibrations exactly double that of another, the second is said to be the octave of the first; and if the number of vibrations in this last be doubled again it is the second octave; and so on up. The range of the ear is about eleven octaves. A note and its octave have such extraordinary likeness to each other that if not sounded in pretty quick succession many, even cultivated ears, fail to detect the difference; and the explanation of the likeness in a psychical point of view is difficult if not impossible. A man and a woman singing the same note in accord seem, to most persons, to be using the same pitch, when in reality they are commonly an octave apart. There are, however, some women's voices pitched as low as a man's. Few persons have any notion of 'absolute pitch,' that is, have the power to tell the pitch of a note sounded apart from any note of known pitch, sounding at or near the same time. It is said that this gift is not possessed by more than one in a hundred, while perhaps about the same ratio of people have no notion of differences in pitch under any circumstances. Such people are music-deaf as some are color-blind.

The complete range of sounds for musical purposes is reckoned at about nine octaves, though something must be cut off the top and bottom for all practical purposes, making the effective range between six and seven octaves; or, in vibrations, between 40 and 4000 to the second. Organ pipes are actually made to embrace this entire interval, that is to say, from two octaves below the lowest note of a bass voice, to about three octaves above C in alt. This gives a difference in length of pipe from $\frac{3}{4}$ of an inch to 32 feet: but an air played in the lowest or highest octave can scarcely if at all be recognized.

Without going into detail, there are six notes in the diatonic scale interpolated between each two octaves with somewhat varying intervals. The intervals between the several notes expressed in the relative number of vibrations are in the scale of C proximately $\frac{9}{8}$, $\frac{5}{4}$, $\frac{4}{3}$, $\frac{3}{2}$, $\frac{5}{3}$, $\frac{15}{8}$.

In man the voice is due to the vocal organ which is placed at the top of the windpipe, the extremity of which is almost closed by two elastic membranes, called the *vocal cords*. These are caused to vibrate by the passage of air from the lungs through the trachea or windpipe, and are made to vary in tension, by a wonderful muscular arrangement, so as to cover a large range in pitch and intensity. Suppose two india-rubber bands to be stretched over the mouth of a fairly large glass tube, leaving a slit between them, and air to be forced through this slit. These edges would be thrown into vibrations of greater or less rapidity according to the tension. This is very like the action of the vocal cords, only remembering that this tension is regulated by one at will. The sweetness and firmness of voice is determined by the smoothness and evenness of the edges of this slit in the glottis, and the accuracy with which they fit together at regular intervals in opening and closing. If the edges are jagged, or strike each other in vibrating, the voice is harsh, or husky. The excellence of the organ will depend upon the rapidity and certainty with which the cords can be stretched or relaxed, the opening enlarged and contracted, and especially upon the form of the cavity of the mouth for sustaining and re-enforcing the sounds which proceed from the glottis.

In men, by the development of the larynx, the vocal cords become much elongated, as compared with those of women, the ratio being as 3 to 2, so that the male voice is lower in pitch and stronger and fuller. A rapid change takes place in the development of the cords in boys at the age of puberty: they become for a time uncertain and squeaky, and generally fall an octave in pitch. There is a considerable increase of the glottis of the girl also, but only about one-half that of the boy, and the voice does not change. In advanced life, or by disease, decided changes take place in the physical conditions of the vocal organs, with a consequent change in voice; but a great deal of the change in most people's voices is due to a lack of attention or care in keeping the vocal cords up to their work.

The range of the human voice is quite variable. The ordinary register is about two octaves, but certain rare voices have a range of three and a half. An extraordinary case is recorded upon the authority of Mozart — that of Lucrezia Ajugari, who gave purely the third octave above middle C, while she trilled freely on the *re* below. Hers was the most remarkable high-pitched voice ever known, — an octave and a half above the ordinary soprano. A basso named Gaspard Forster passed from the *fa* of the third octave below middle C to *la* above, the lower end of his register being a full octave below that of the ordinary bass.

Helmholtz has shown that the conformation of the cavity of the mouth acts as a resonator, and so has very much to do with the volume and elasticity of the voice, and the original structure of the whole vocal apparatus determines the power and smoothness of one's vocalization in speech and song; but a large margin is still left in which one's own effort can modify favorably or the reverse; and commonly, in this country, it is for the worse. We may as well confess that the American voice, in speech, as a rule, is the worst in the world. The one wretched fault is nasality. This is not the work of nature, but is the result of volitional neglect, or falsely directed effort. There is scarcely one person in a hundred in this country who has not made his or her voice thin and nasal, by not using the right muscles in regulating the tension of the vocal cords, driving out of the voice the lower overtones which give it richness, volume, and fulness.

CHAPTER VII.

THE SENSE OF VISION.

Mechanism of the eye. Structure of retina. End-organs. Rods and cones. Mechanical basis of vision. Color. 'Consecutive' images. Tone, intensity, saturation. Yellow spot.

THE eye is, if possible, more wonderful in its mechanism than the ear. It is primarily a complete optical instrument, and, as such, sufficiently wonderful; but the refinement of mechanical contrivance is found, not in the optical arrangement but in the preparation of light, by the end-organs in the retina to be transmitted to the brain.

In the optical arrangement of the instrument there are the following transparent media — the 'cornea,' the 'aqueous humor,' the 'crystalline lens,' and the 'vitreous humor.' The cornea is the outer, horny covering of the eyeball which is transparent, and the outer and inner surfaces being parallel, it exerts no effect in refracting or bending the rays of light. All the other media do exert such power; but by far the most considerable effect in this way is produced by the 'crystalline lens.' This is placed just behind the diaphragm of the iris which automatically regulates the admission of light by enlarging or contracting the pupil. The crystalline is a bi-convex lens with its axis on a line with the centre of the pupil, and covering in extent the pupil and iris. Its front surface is not so much curved as the posterior surface, and by the arrangement of certain delicate muscles, the front surface is made to bow more or less in order to accommodate the eye to the rays of light from objects at different distances. The posterior surface remains fixed in curvature. It is not homogeneous in structure, but composed of layers like an onion. This is an arrangement of Nature to correct for chromatic and spherical aberration.

From the cornea to the posterior surface of the lens the distance is about one-third of the optical axis. All the remaining portion of the interior of the eyeball is filled with the 'vitreous humor,' which is as translucent as glass, and jelly-like in consistency. It does not seem to have any very great significance physiologically or optically, but to be intended rather to hold everything compact within the eyeball.

Next the vitreous humor comes the 'retina,' which lines the whole of the back part of the eyeball, extending about two-thirds of the way towards the front. Behind the retina comes the 'choroid coat' which is the middle one of the three envelopes, or tunics, of the eyeball. It is quite dark, inclining to black, and extends far forward, joining on to the iris in front. It is abundantly supplied with blood-vessels and nerves. In Albinos, and in many mammals also, it contains no pigment, though the structure is the same; which gives their eyes a peculiar iridescent lustre. In the horse, and in ruminant animals this lustre of the eye is also seen, but it is due to the reflection of bundles of tissue. In cats it is due, according to Schultze, to cells containing double refractory crystals. When this coat is dark a part of the light which enters the eye is absorbed, and the pupil is black.

It has been suggested that in those animals presenting an iridescence the eye is probably more sensitive to light of feeble intensity. The iris, which is really a continuation of the choroid coat, is too well known to require description.

Outside of all is the 'sclerotic coat,' which with the cornea forms a complete protective covering of the entire instrument, except where it is pierced by the optic nerve, shortly to be noticed. It is a firm, unyielding, fibrous membrane, white, except the cornea, which is transparent. To it are attached the muscles for the movement of the eyeball.

There are many accessory arrangements of the most delicate nature for the protection and management of the eye, such as eyelids, eyebrows, lachrymal apparatus, etc., which in this mere outline we need not stop to consider. To understand the eye in its marvellous perfections would be a study in itself.

All that part of the eye in front of the retina is simply an optical arrangement by which an image is produced upon the retina, such as may be accomplished by any bi-convex lens, only infinitely more perfect. The image is inverted, as must be the case, with such a lens.

In the retina are placed the end-organs of vision, and their natural stimulus is light. There is some question as to whether they can be excited by mechanical or electrical means. Such stimuli do without doubt produce luminous impressions when applied either to the optic nerve or the eyeball, but it is doubtful whether the retina itself is affected by any stimulus but light. It is thought that there is some chemical action on the retina, but what, is not yet well ascertained.

The optic nerve pierces the sclerotic coat and enters the retina about $\frac{1}{6}$ of an inch to the inner or nasal side of the optical axis. The diameter of the optic nerve where it pierces the retina is about $\frac{1}{12}$ of an inch, varying somewhat in different eyes, and since at the point of entrance the nerve is wholly insensible to light stimulus, it forms what is called the 'blind-spot.' By a little contrivance any one can easily discover it in his own eye. Exactly in the centre of the retina—the place of most perfect optical effect, is the 'yellow spot,' with a diameter between $\frac{1}{18}$ and $\frac{1}{20}$ of an inch. This spot is best developed in man, and apes among mammals, though it has been shown to exist in reptiles.

The retina is a highly complex structure, colorless and translucent, very soft, composed of numberless cells, fibres, end-organs, connective tissues and blood-vessels, arranged according to Max Schultze,—a high authority,—in ten layers, including the inner pigment cells of the choroid. It is not necessary to go into details, but beginning at the delicate membrane which forms the inner lining of the retina, there lie immediately behind, and parallel to it, the nerve-fibres, raying out in every direction from the optic nerve. They surround, but do not cover, the yellow spot, where they are thickest, gradually thinning out towards the edges of the retina.

The several layers which follow, from within outward, are a mass of cells or fibrils of various structure, but their functions are not certainly known. It is agreed, however, that the true end-apparatus of vision is found in the ninth, or layer next the choroid coat, called the rod and cone layer, which is composed of " multitudes of elongated bodies arranged side by side like rows of palisades, and vertically to the surfaces of the retina." They are of two kinds, some of them cylindrical, and called the 'rods' of the retina, and others conical or flask-shaped, and called the 'cones' of the retina. The rods are about $\frac{1}{350}$ of an inch in length, and the cones something like half as long, the diameters being about $\frac{1}{10000}$ of an inch for the cones and $\frac{1}{14000}$ for the rods. On the exterior of this layer of rods and cones lies the pigment-epithelium of the choroid, a perfect mosaic of hexagonal cells, in closest connection with it, sending up pigmented processes between the rods and cones.

There is no doubt among physiologists that the true nervous effect which gives rise to vision, is due to these rods and cones of the retina; and that the vibrations of the luminiferous ether pass through the inner layers and reach these end-organs, where the nervous process really begins.

The mechanical basis of vision, as indeed of all sensation, is motion; but in the case of sight it is infinitely refined as compared with that of hearing — the sense in which vibratory motion is most clearly demonstrable to touch and sight. The highest possible sensation of the ear corresponds with a number of vibrations many million times less than that of the lowest possible sensations of the eye.

The fact is, science makes large demands upon our credulity in the phenomena of light. In the first place, it is compelled to assume the existence of a substance — the luminiferous ether, — pervading all space, for the transmission of light at all; and the characteristics of this substance which the undulatory theory of light force upon it, are marvellous in the extreme. It has to be so subtle as to allow all bodies to pass through it, or itself to pass through them, without the possibility of discovering that it exists at all as a resisting medium, and yet it has to be a solid of a rigidity immensely greater than the hardest substance of the earth. It is the vibratory motion of this substance which by its action on the retina gives rise to the sensation of light.

Sir Isaac Newton discovered that common light, that is, white light, as that of the sun, is not simple but composite. As everybody knows, a ray of sunlight passed through a prism gives all the colors of the rainbow, and this succession of colors arranged as given by a prism, forms what is called the solar spectrum, with red at one end and violet at the other. The red end of the spectrum is the lowest order of luminosity, and is physically due to the least rapid vibratory motion of the luminiferous ether, or, which means the same thing, that order of vibrations which has the greatest wave length. In dark red light the number of vibrations in a second is 392,000,000,000,000 and the wave length 0.000760 of a millimetre. At the other extreme end of the spectrum is the violet with 757,000,000,000,000 per second and having a wave length 0.000397 of a millimetre. These limits give the extreme range of the eye, and between them lie all the other colors.

The space covered by the prismatic colors in the spectrum embraces those ethereal vibrations which lie between the limits already given, but there are less rapid vibrations below the lower end, and more rapid, far above the upper end. Those below produce the most powerful heat effects, and are called 'heat rays,' while those above are most active in producing chemical effects, and are called 'actinic rays.' That these phenomena are all connected may be readily made to appear. If a current of electricity be passed through a platinum wire the temperature will gradually rise as the intensity of the current is increased, until, at a temperature of about 540°, it will begin to glow. The light first emitted will be red; then will be added yellow, green, blue, and violet in succession. When it reaches white heat it emits all the prismatic colors.

The number of colors is really infinite, since they grade into each other imperceptibly, but there are three colors which seem to be fundamentally different from each other, and from all others. They are the red, green, and violet. All other colors are composite and can be produced by mixing these three, two and two, in varying proportions; but neither the red, green, or violet can be so produced. They are therefore primary, and it is thought that they correspond to three specific activities of the rods and cones of the retina. They may be regarded as the three fundamental sensations of the eye. Such is the theory of Dr. Young, elaborated by Helmholtz. Homogeneous or monochromatic light excites all three, but with varying intensities according to the length of the wave. Long waves excite most strongly the red, medium waves the green, and the shortest, violet. The spectrum presents a succession of color-bands, quite distinct though they grade into each other; they are called the 'prismatic colors'—red, orange, yellow, green, cyan-blue, ultramarine-blue, and violet. The bands are not of equal breadth, the blue being greatest in extent. If these be arranged on a circle with purple between the violet and red, and with subdivisions as indicated in the figure here given, the two colors lying at the extremities of any diameter when mixed will produce white. In any set of such colors, either is called 'complementary' of the other. Any color lying on the circumference between the red and green can be produced by mixing these two in proper proportions. So also with those between green and violet. Purple, which is not a prismatic color, is produced by red and violet. All the colors of the spectrum together, of course, produce white.

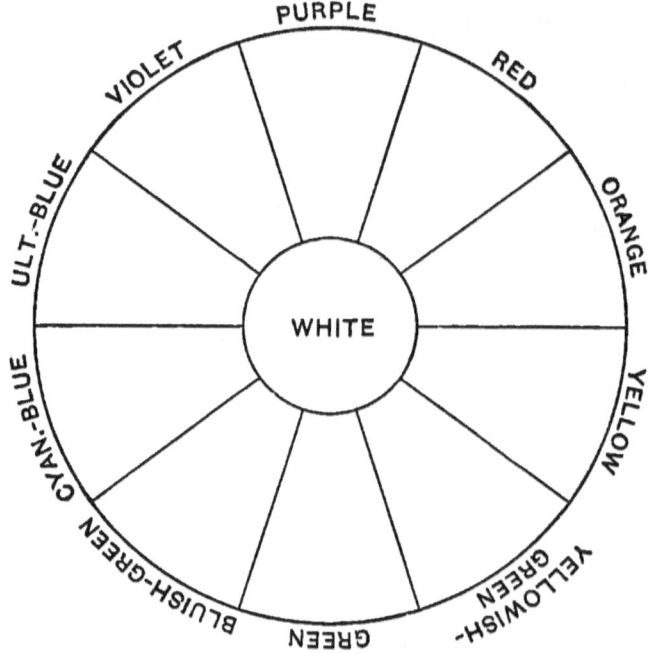

Mixing colors, however, is not the same thing as mixing pigments, — the results are quite different. For example, if chrome-yellow and ult-blue be mixed, a green is the result; but if with the same pigment a disk be painted something over half of it blue, and the rest yellow, and then made to revolve, the result will be white; or rather an approximation. In this last case the mixture takes place in the retina itself, by reason of what is called 'persistence of retinal impressions.' When the retina is once excited the impression lasts for a short time — from $\frac{1}{50}$ to $\frac{1}{80}$ of a second — after the removal of the stimulus.

If one looks fixedly at a bright light for a moment, and then quickly closes the eye, one sees a luminous image of the object, — for a short time quite bright, but gradually fading. Many curious illusions are produced through this principle of persistence.

These 'accidental' or 'consecutive' images are of two kinds, —'positive,' where the eye being fixed but a short time, say, $\frac{1}{8}$ of a second, the image has its lights and shades in the same order as in the object; — 'negative,' where the gaze having been prolonged, the image has its bright parts where the dark parts of the object were, and the reverse. If the gaze be upon, say, a disk of yellow paper on a gray ground for fifteen or twenty seconds, and the disk be suddenly removed without disturbing the eye, there will appear in its place, a blue image, that is, the complementary color will appear; and so with any other color. This is accounted for physiologically by the theory that the retina becomes fatigued by the continued stimulation, and upon the removal of the stimulus, the eye does not respond to the low stimulation of that color in the gray, but readily to its complementary color which is also there. It is not improbable that some chemical activity also enters the problem.

Colors have three special characteristics, 'Tone,' 'Intensity,' and 'Saturation.' Tone depends upon the number of vibrations per second, and we distinguish different color tones as we pass from the red end to the violet end of the spectrum, but without marked lines of separation. Intensity is doubtless due to the amplitude of the vibrations, and in sensation means the greater or less brightness. A tone is said to be 'saturated' or 'pure' when there is no white light mixed with it. It is difficult to get perfectly saturated tones except in the spectrum. There are an infinite number of 'tints' or 'shades' which result from mixtures of colors.

The retina is not all equally sensitive. The 'yellow spot' is altogether the most sensitive, and it is doubtful whether distinct vision is possible except when the image is made to fall on this spot. This is why the eye is in such constant motion, and an extremely small movement, when the eye is looking in the direction of an object, brings the image on this spot. Though the spot is so small it corresponds to a visual angle of from 2° to 4°. Its extreme sensibility is due no doubt to the immense number of cones — said to be one million in an area not greater than $\frac{1}{100}$ of a square inch.

CHAPTER VIII.

CHASM BETWEEN MECHANISM AND CONSCIOUSNESS.

Physiological research with respect to psycho-mechanisms. Protoplasm not pure and simple matter. Professor Romanes quoted. Darwin, Huxley, Tyndall, and Spencer not materialists. Hobbes quoted. The problem of relation between physiology and consciousness. The chasm recognized. Leaders of science quoted.

WE have now before us in outline, substantially all that the latest physiological research can tell us touching the human mechanism in its relation to the psychic powers. To one without some philosophical training, the case might well seem to be closed and consciousness and thought accounted for. It is not too much to say, perhaps, that the majority of those engaged in physical research are quite satisfied that matter is the cause of mind; and it is to be feared that immense numbers of people are quietly acquiescing in this unscientific science. It is unscientific because science itself, by the voice of its chosen leaders, declares it to be so.

But before we hear the physicists speak, let us look at the question a moment for ourselves in the light of common sense. In the first place the whole physiological structure is built up out of protoplasm; but protoplasm is not pure and simple matter. It is vitalized matter, and so vastly different from dead or inert matter. This life-factor cannot be produced from dead matter, so far as known, notwithstanding persistent efforts to that end. The question of 'spontaneous generation' may be considered as set at rest, since all efforts to evolve life from dead matter have failed; and the scientific world has substantially given it up. Since it is impossible to prove a negation by experiment, this question can never be settled beyond contradiction; but we are bound to take the fact as it stands, and recognize in protoplasm that very something, vitality, which is not a necessary factor of matter. This non-material factor confessedly gives to matter the potentialities through which organization and co-ordination are accomplished in all animate structures; and it is through this factor that that one indisputable fact of the universe which we call 'personality' is made conceivable. It is thus absolutely necessary for the physicist to start with matter *plus* the one fact through which explanation becomes possible, or through which anything ever obtains which needs explanation. This the leaders of scientific thought clearly see, and have explicitly stated, as will abundantly appear; but the metaphysical and theological world, as well as most of the popular-science luminaries, are either in ignorance of the fact, or regard it as of no importance; and hence a needless alarm on the one hand, and a scornful satisfaction on the other.

Or, again, let us go back to the ultimate basis of certitude, which we found in the beginning to be the conscious ego — the self — the source and centre of all knowing. We saw that we could doubt of all things whatever, protoplasm, the whole nervous system, motion, the brain and all its functions, but the thinker cannot doubt that he thinks, — he cannot doubt that he thinks he has, or is, a personality. Protoplasm, or matter, or whatever else there may be, then, has no better warrant than thought for the conviction, that it is what it is, or, is at all.

Let me use once more the words of Professor Romanes: "All our knowledge of motion, and so of matter, is merely a knowledge of the modification of mind. That is to say, all our knowledge of the external world, including the knowledge of our own brains, is merely a knowledge of our own mental states. Let it be observed that we do not even require to go as far as the irrefutable position of Berkeley, that the existence of an external world without the medium of mind, or of being without knowing, is inconceivable. It is enough to take our stand on a lower level of abstraction, and to say that whether or not an external world can exist in any absolute or conceivable sense, at any rate it cannot do so *for us*. We cannot think any of the facts of external nature without presupposing the existence of a mind which thinks them; and therefore, so far at least as we are concerned, mind is of necessity prior to anything else. It is for us the only mode of existence which is real in its own right; and to it, as a standard, all other modes of existence which may be *in*ferred must be *re*ferred. Therefore, if we say that mind is a function of motion, we only say in a somewhat confused terminology, that mind is a function of itself."

Philosophy has always had to grapple with this problem of the possible nexus between mind and matter. Not to go back beyond Descartes, a number of theories have been proposed to solve the difficulty, all, now, rather curious than useful, but we shall leave them on one side for the present. The only theory which concerns us here is that of pure materialism, but that is hardly worth consideration, since it is doubtful whether there be any materialists, in the proper sense, anywhere. It is certain that Darwin was not, and that Huxley, Tyndall, Spencer, and all that school who understand their leaders are not, however unduly they may emphasize the mere physical side. One cannot be a materialist, — one, that is, who is sufficiently acquainted with the subject to be entitled to an opinion, since we do not even know that there is any such thing as matter, except through the necessary postulate that mind is: and we cannot conclude that mind is but the product of matter, since the effect to be accounted for, in such case, is necessarily the factor through which this, or any other cause is demanded. Professor Romanes puts this strongly:

"'Motion,' says Hobbes, 'produceth nothing but motion';

and yet he immediately proceeds to assume that in the case of the brain, it produces not only motion, but mind. He was perfectly right in saying that, with respect to its movements, the animal body resembles an engine or a watch; and if he had been acquainted with higher evolution in watch-making, he might with full propriety have argued, for instance, that in the compensation balance, whereby a watch adjusts its own movements in adaptation to external changes of temperature, the watch is exhibiting the mechanical aspect of volition. And, similarly, it is perhaps possible to conceive that the principles of mechanism might be more and more extended in their effects, until, in so marvellously perfect a structure as the human brain, all the voluntary movements of the body might be originated in the same mechanical manner as the compensating movements of a watch; for this, indeed, as we have seen, is no more than happens in all the nerve-centres other than the cerebral hemispheres. If this were so, motion would be producing nothing but motion, and upon the subject of brain-action there would be nothing further to say. Without consciousness I should be delivering this lecture; without consciousness you would be hearing it; and all the busy brains in this University would be conducting their researches, or preparing for their examinations, mindlessly. Strange as such a state of things might be, still motion would be producing nothing but motion; and, therefore, if there were any mind to contemplate the facts, it would encounter no philosophical paradox: it would merely have to conclude that such were the astonishing possibilities of mechanism. But, as the facts actually stand, we find that this is not the case. We find, indeed, that up to a certain level of complexity mechanism alone is able to perform all the compensations of adjustment which are performed by the animal body; but we also find, that beyond this level such compensations or adjustments are never performed without the intervention of consciousness. Therefore the theory of automatism has to meet the unanswerable question — How is it that in the machinery of the brain motion produces this something which is not motion? Science has now definitely proved the correlation of all the forces: and this means that if any kind of motion could produce anything else that is not motion, it would be producing that which science would be bound to regard as in the strictest sense of the word a miracle. Therefore, if we are to take our stand upon science, and this is what materialism professes to do, — we are logically bound to conclude, not merely that the evidence of causation from body to mind is not so cogent as that of causation in any other case, but that in this particular case causation may be proved again in the strictest sense of the term a physical impossibility."

Professor M'Kendrick, University of Glasgow, says: "No one doubts that consciousness has an anatomical substratum, but the great problem of the relation between the two is as far from solution as in the days when little or nothing was known of the physiology of the nervous system. Consciousness has been driven step by step upwards until now it takes refuge in a few thousand nerve-cells in a portion of the gray matter of the cortex of the brain. The ancients believed that the body participated in the feelings of the mind, and that, in a real sense, the heart might be torn by contending emotions. As science advanced, consciousness took refuge in the brain, first in the medulla, and lastly in the cortex. But even supposing that we are ultimately able to understand all the phenomena — chemical, physical, and psychological, of this intricate ganglionic mechanism, we shall be no nearer a solution of the problem of the connection between the objective and subjective aspects of the phenomena. It is no solution to resolve a statement of the phenomena into mental terms or expressions, and to be content with pure idealism; nor is it any better to resolve all the phenomena of mind into terms describing physical conditions, as in pure materialism. A philosophy which recognizes both sets of phenomena, mutually adjusted and ever interlacing, may be no explanation; but at all events it is unpretentious, recognizes the facts, and does not delude the mind by offering a solution which is no solution at all."

As we have said, this point is so clearly seen by all schools of thought, that it can hardly be said that there are any materialists in the proper sense of the word at the present time.

This is a matter of such moment, that it will be well to substantiate the statement by quotations at some length from a few of the most distinguished leaders of scientific thought; and we begin with Dr. Tyndall.

He says, in his lecture on "Matter and Force": "While I as a man of science feel a natural pride in scientific achievements, while I regard science as the most powerful instrument of intellectual culture, as well as the most powerful ministrant to the material wants of men; if you ask me whether science has solved, or is likely in our day to solve, the problem of this universe, I must shake my head in doubt. . . . As far as I can see, there is no quality in the human intellect which is fit to be applied to the solution of the problem. It entirely transcends us. The mind of man may be compared to a musical instrument with a certain range of notes, beyond which in both directions we have an infinitude of silence. The phenomena of matter and force lie within our intellectual range, and as far as they reach we will at all hazards push our inquiries. But behind, and above, and around all, the real mystery of this universe lies unsolved, and as far as we are concerned is incapable of solution."

In his lecture on the "Scope and Limits of Scientific Materialism," he is, if possible, even more pronounced. He says: "I am not mincing matters, but avowing nakedly what many scientific thinkers more or less distinctly believe. The formation of a crystal, a plant, or an animal is in their eyes a purely mechanical problem, which differs from the problems of ordinary mechanism in the smallness of the masses and the complexity of the processes involved. Here you have one half of our dual truth; let us now glance at the other half. Associated with this wonderful mechanism of the animal body we have phenomena no less certain than those of physics, but between which and the mechanism we discern no necessary connection. A man, for example, can say, *I feel, I think, I love;* but how does *consciousness* infuse itself into the problem? The human brain is said to be the organ of thought and feeling; when we are hurt the brain feels it, when we ponder it is the brain that thinks, when our passions or affections are excited it is through the instrumentality of the brain. . . .

"The relation of physics to consciousness being thus invariable, it follows that, given the state of the brain, the corresponding thought or feeling might be inferred; or given the thought or feeling, the corresponding state of the brain might be inferred. But how inferred? It would be at bottom not a case of logical inference at all, but of empirical association. You may reply that many of the inferences of science are of this character; the inference, for example, that an electric current of a given direction will deflect a magnetic needle in a definite way; but the case differs in this, that the passage from the current to the needle, if not demonstrable, is thinkable, and that we entertain no doubt as to the final mechanical solution of the problem. But the passage from the physics of the brain to the corresponding facts of consciousness is unthinkable. Granted that a definite thought, and a definite molecular action in the brain, occur simultaneously; we do not possess the intellectual organ, nor apparently any rudiment of the organ, which would enable us to pass, by a process of reasoning, from the one to the other. They appear together, but we do not know why. Were our minds and senses so expanded, strengthened, and illuminated as to enable us to see and feel the very molecules of the brain; were we capable of following all their motions, all the groupings, all the electric discharges, if such there be; and were we intimately acquainted with the corresponding states of thought and feeling, we should be as far as ever from the solution of the problem, 'How are these physical processes connected with the facts of consciousness?' The chasm between the two classes of phenomena would still remain intellectually impassable. Let the consciousness of '*love*,' for example, be associated with a right-handed spiral motion of the molecules of the brain, and the consciousness of *hate* a left-handed spiral motion. We should then know when we love that the motion is in one direction, and when we hate that the motion is in the other; but the '*why?*' would remain as unanswerable as before.

"In affirming that the growth of the body is mechanical, and that thought as exercised by us, has its correlation in the physics of the brain, I think the position of the 'materialist' is stated, as far as that position is a tenable one. I think the materialist will be able finally to maintain this position against all attacks; but I do not think, in the present condition of the human mind, that he can pass beyond this position. I do not think he is entitled to say that his molecular groupings and his molecular motions *explain* everything. In reality they explain nothing. The utmost he can affirm is the association of two classes of phenomena, of whose real bond of union he is in absolute ignorance. The problem of the connection of body and soul is as insoluble in its modern form as it was in the pre-scientific ages."

Professor Tyndall goes even further than this: in the October number of the *Contemporary Review*, 1872, he says: "It is no departure from scientific method to place behind natural phenomena a Universal Father, who, in answer to the prayers of his children, changes the currents of phenomena. Thus far theology and science go hand in hand."

Again, at Manchester he declares: "I have, not sometimes

but often, in the springtime . . . observed the general joy of opening life in nature, and I have asked myself the question, Can it be that there is no being in nature that knows more about these things than I do? Do I, in my ignorance, represent the highest knowledge of these things existing in the Universe? Ladies and Gentlemen, the man that puts that question fairly to himself, if he be not a shallow man, if he be a man capable of being penetrated by profound thought, will never answer the question by professing that creed of Atheism which has been so lightly attributed to me."

Professor Huxley is not in the least behind Professor Tyndall in the clearness with which he sees the limits of mechanical action, nor less pronounced in declarations. He says, in his " Lay-Sermon " on " The Educational Value of the Natural History Sciences " : " What is the cause of this wonderful difference between the dead particle and the living particle of matter appearing in other respects identical? That difference to which we give the name of Life? I, for one, cannot tell you. It may be that, by and by, philosophers will discover some higher laws of which the facts of life are particular cases — very possibly they will find out some bond between physico-chemical phenomena on the one hand, and vital phenomena on the other. At present, however, we assuredly know of none ; and I think we shall exercise a wise humility in confessing that, for us at least, this successive assumption of different states — (external conditions remaining the same)— this *spontaneity of action* — if I may use a term which implies more than I would be answerable for — which constitutes so vast and plain a practical distinction between living bodies and those which do not live, is an ultimate fact ; indicating as such, the existence of a broad line of demarcation between the subject-matter of Biological, and that of all other sciences."

In his address on the " Physical Basis of Life," he says : " Past experience leads me to be tolerably certain that, when the propositions I have just placed before you are accessible to public comment and criticism, they will be condemned by many zealous persons, and perhaps by some few of the wise and thoughtful. I should not wonder if 'gross and brutal materialism' were the mildest phrase applied to them in certain quarters. And, most undoubtedly, the terms of the propositions are distinctly materialistic. Nevertheless two things are certain; the one, that I hold the statements to be substantially true; the other, that I, individually, am no materialist, but on the contrary, believe materialism to involve grave philosophical errors. This union of materialistic terminology with the repudiation of materialistic philosophy, I share with some of the most thoughtful men with whom I am acquainted."

Again he says: "All who are competent to express an opinion on the subject are at present agreed that the manifold varieties of animal and vegetable life have not either come into existence by chance, nor result from capricious exertions of creative power, but that they have taken place in a definite order, the statement of which order is what men of science term natural law."

But he reaches the highest possible pitch in the following energetic expression: "How it is that anything so remarkable as a state of consciousness comes about as the result of irritating nervous tissue, is just as unaccountable as the appearance of the Djinn when Aladdin rubbed his lamp."

The following assertion which he makes in speaking of 'Matter and Force' is all that metaphysic can demand: "It is an indisputable truth that what we call the material world is only known to us under the forms of the ideal world."

He says again, speaking of physics and metaphysics: "Their differences are complementary, not antagonistic, and thought will never be completely fruitful until the one unites with the other."

Herbert Spencer is equally definite in his statements. He says in the *Nineteenth Century* of January, 1884: "Those who think that science is dissipating religious beliefs and sentiments seem unaware that whatever of mystery is taken from the old interpretation is added to the new. Or rather, we may say, that transference from one to the other is accompanied by increase, since for an explanation which has a seeming feasibility, science substitutes an explanation which, carrying us back only a certain distance, there leaves us in the presence of the avowedly inexplicable. . . . But amid the mysteries which become the more mysterious the more they are thought about, there will remain the one absolute certainty that he is ever in the presence of an Infinite and Eternal energy." To this Infinite and Eternal energy from which "all things proceed," he only hesitates to apply the word 'Person,' because "though the attributes of personality, as we know it, cannot be conceived by us as attributes of the Unknown Cause of things, yet duty requires us neither to affirm nor deny personality; but the choice is not between personality and something lower, but between personality and something higher, and the ultimate power is no more representable in terms of human consciousness than human consciousness is representable in terms of plant functions." He says further: "I held at the outset, and continue to hold that the Inscrutable Existence, which science in the last resort is compelled to recognize as unreached by its deepest analysis of matter, motion, thought, and feeling, stands towards our general conception of things in substantially the same relation as does the Creative Power asserted by Theology."

Darwin bears testimony to the same necessity of postulating an ultimate Power which is not 'blind.' He says in the "Descent of Man": "I am aware that the conclusions arrived at in this work will be denounced by some as highly irreligious; but he who thus denounces them is bound to show why it is more irreligious to explain the origin of man, as a distinct species, by descent from some lower form, through the laws of variation and natural selection, than to explain the birth of the individual through the laws of ordinary reproduction. The birth of the species and of the individual are equally parts of that grand sequence of events which our minds refuse to accept as the result of blind chance."

Dr. Maudsley, in that direful, pessimistic book, "Body and Will," speaks upon this point as follows: "Is there any good reason why the doctrine of *evolution* and the doctrine of *epigenesis* should be opposed to one another as irreconcilable doctrines? More correctly perhaps epigenesis is an event of evolution, and evolution impossible without epigenesis; for evolution, strictly speaking, is the unfolding of that which lies as a pre-formation in germ, which a new product with new properties manifestly does not, any more than the differential calculus lies in a primeval atom; while epigenesis signifies a state which is the basis of, and the causative impulse to, a new and more complex state. There is a leap; and it is not good philosophy to blindfold ourselves with a big word when taking the leap, as some evolutionists will have us do, and then protest that we have not taken it."

The great German physiologist, Du Bois Reymond, is just as pronounced as any of these distinguished Englishmen. He says, "If we had an absolutely perfect knowledge of the body, including the brain and all the changes in it, the psychical state called sensation would be as incomprehensible as ever. For the very highest knowledge we could get would reveal only matter in motion, and the connection between any motions of any atoms in my brain and such unique undeniable facts as that I feel pain, smell a rose, or see red is thoroughly incomprehensible."

M. Pasteur, upon taking his place in the French Academy must have astonished that body of savants as he uttered the following: "Beyond the starry vault above us, what is there? Other starry skies. Well and beyond those? The human mind, swayed by an invincible impulse, will never cease to inquire what there is beyond: and there is no point in time and space which can set at rest the implacable question. It is no use to reply that beyond any given point there is boundless space, time, or magnitude. Such words convey no tangible meaning to the human mind. The man who proclaims the existence of the Infinite (and there is no man who does not) accumulates in that bare statement more supernatural elements than are to be found in the miracles of all religions; for the notion of the Infinite has this double character — that it is at once self-evident — that it forces itself upon the mind, and yet is incomprehensible. . . . This positive and primordial notion with all its consequences in the life of societies, positivism sets at naught. The Greeks understood the power of the unseen world. They have left us the noblest word in our language 'enthusiasm,' ἐν θεός — an inner God. The greatness of human deeds can be measured by the inspiration that gives them birth. Happy the man who has an inner God, — an ideal of beauty, — and who obeys his behests. The ideal of art, the ideal of science, the ideal of country, the ideal of the verities of the Gospel — those are the living sources of great ideas and noble deeds — they are illuminated by a gleam from the Infinite."

Like quotations could be accumulated from any number of eminent physiologists, as well in America as in Europe, but there is no need. It can hardly be disputed that science does not countenance a pure, bald atheistic materialism, but on the contrary maintains an ultimate power behind all physical phenomena.

CHAPTER IX.

PERSONALITY IN ITS PSYCHICAL ASPECT.

Analysis of the psychical factor of personality. Three fundamental modes of the self — sensation, cognition, and conation. A tri-unity, inseparable but logically distinguishable. Sub-consciousness. Unity and plurality.

THE testimony, as we have seen, of the leaders of science to the existence of a chasm in thought, between all possible mechanisms, and that ineffable somewhat which we know at first hand, and name Personality, is as unqualified as the most rigid metaphysician or theologian could ask. Let us now consider what is involved in personality as we know it in consciousness.

Take any, — the simplest act of experience. I look at my watch and note the hour. A little reflection will reveal to us three several modes, or psychical phases of the conscious personality in this simple event. First, the figures and hands on the face of the watch act as stimuli to the organ of sight and produce in me, somehow, the sensation of an object without, variously marked: second, I take note of the relative positions of the hands with respect to the figures, and understand the hour indicated: and third, I am conscious of effort or attention throughout the whole event. These three facts of the conscious personality are called, respectively, 'sensibility,' 'cognition,' and 'conation,' — which, in a general way, answer to what are commonly known as feeling, thought, and will.

The one element common to all three of these fundamental modes of the self is consciousness, which may be called the daylight of personality. It does not itself see, — it does not itself feel, modify or arrange, but it lights up the psychical world, and thus is the occasion and not the cause of all these.

While we must clearly distinguish the three elementary modes of the self from each other, they cannot be made to stand apart as numerically separate, and so are not three totally different kinds of knowledge. Neither can they be confused, or made to pass one into the other. Each, while not being either of the other two, nor in any conceivable way like either of them, presupposes both; and there can never be any, the simplest act of illuminated or conscious knowledge in which all three are not found as constituent factors. These are simply facts of our psychical nature which must be admitted by all who understand the force of the terms in which they are enunciated.

We have already seen something of what is known, since Herbart, as the 'threshold value' of sensation; that is, of the intensity which sensation must reach before it, so to speak, flows over the threshold of consciousness and becomes distinguishable in the self. It is well established that there are abundant sensations, or grades of stimulation of the sense-organs which are not and cannot be reached by the illuminating power of consciousness, and yet largely affect the personality. These all lie in the region which has been happily called sub-consciousness. In the nature of the case it is a region, not of knowledge, but of personal activities which make it possible for the higher, conscious activities to exist. We may pass it by, however, for the present, since we are considering only what we know in consciousness.

To resume: let us try to see how each of the three primary modes of the conscious self presupposes the other two. The result of a stimulus cannot be a sensation in consciousness unless it be known, and it cannot be known unless it come within the sphere of attention. But to know is to understand or think, and to attend is to conate or energize, — actively that is, or with purpose, or passively, that is, in response to solicitation. Any — the simplest effort, must fall within the general domain of the will.

But even after we rise into the clearly marked region of the purposive will, we see that external stimuli, and the reactions of the physical factor, do not necessarily result in differentiated sensation, though far above the threshold value. The greater part of what goes on about us is quite lost to consciousness because we do not attend. When occupied we do not hear the ticking of the clock. The miller is not conscious of the whir and jar of his machinery, and the mother is too often wholly oblivious to the din and riot of her progeny. One may be seriously hurt and know nothing about it. Soldiers wounded in action often get the first intimation of it from their comrades. Rapt attention, intense interest, passion or fear constantly renders one unconscious of what is going on about one when out of the focus of attention.

That there could be no thought without sensation is sufficiently obvious. The self in a state of absolute isolation from the beginning, — if such a thing is conceivable, — could have no material of thought and would remain forever without knowledge of any sort.

So with the will. No purposive nor passive activity of the conative power is possible without sensibility and cognition.

And yet, on the other hand, no one of these three modes can be in any wise construed in terms of either of the others. Feeling has a purely subjective quality, *i.e.* its content is wholly a state or condition of the self, not known or knowable to another personality; while thought is in some sort objective, discovering relations which would appear equally to any intelligence in possession of the facts. Feeling is individual; thought is universal. The content of will is effort, with movement as its end. In a rough way, if I may so say, sensation is the object upon which a telescope is directed; thought is the instrument itself, and will is the person using it; while consciousness is the light. But an illustration 'must not be made to go upon all-fours'; and so this must not be taken for more than to emphasize the several phases of any psychic event. It wholly breaks down when the unity of any such event is taken into account.

But all this will come up as we proceed, and so I leave it for the present, with the remark that personality is a mystical tri-unity — a fundamental exemplification of the problem about which philosophy, ancient and modern, has ever busied itself — the co-existence of the 'one and the many,' perhaps a living type of the fundamental mystery of the Christian Faith.

CHAPTER X.

DEVELOPMENT OF THE PSYCHICAL ASPECT OF PERSONALITY.

The relation of the mechanical and psychical factors. Mutually necessary. The human organism at birth. The line between elementary consciousness and self-realization shadowy. Automatic action. Basic-personality. Evolution. Continuity and discontinuity. Instincts. Jelly-specks. Ants. *Chætodon rostratus.* The beaver. Domestic animals. Inverse order of intelligence and instincts. A *de*volution as well as an evolution. Instincts gradually replaced in ascending order of nature.

SENSATION is the common ground upon which the self and the non-self come together. The psychical factor of the personality must be present, or, no matter what character or variety of stimuli, acting on the mechanism, produce in it divers states or conditions, there would be nothing to read or interpret these states, and so no sensation: the non-self must be present, or, no matter how responsive the psychical factor may be, there would be nothing to know or feel, and so no sensation. Of these two factors, the external stimuli may be regarded as prior in time, since the psychical factor waits to be acted on; but logically the psychical factor must be prior, since it must be ready and in waiting. But sensation given, it is impossible to conceive of a total separation, and so there is no actual priority, any more than in an explosion it can be said that the spark is prior to the gunpowder, or the gunpowder prior to the spark. It is a case of 'action and reaction,' which, by a fundamental law of mechanics, are always equal and contrary, carrying with them the necessary notion of simultaneity. An external world out of and apart from all sensation cannot be conceived. The self in a state of absolute isolation, even if sensation had been previously awakened, could not retain sensation, since its own modes or states, in such case, must be regarded as objective, and so take their place with the non-self. We must, therefore, hold fast by the reality of the ego and the non-ego, as we saw in the beginning; but the positive factor is the ego which alone gives meaning or conceivable existence to the negative factor — the non-ego.

It cannot be denied that the child comes into the world in an advanced state of reflex activity. The organism is complete, — muscles, nervous system, and all manner of tissue in full working order. It is impossible to say whether there is any degree of consciousness present at and before birth, or not. It can hardly be denied from and after that event, but there seems to be no sufficient physiological reason to fix upon any exact moment at which it makes its advent, and so no reason why some low order of consciousness may not exist previous to birth. That the self is construed or differentiated in thought, cannot, of course, be contended; but personality must undoubtedly obtain. The self must *be* before it can *know* itself to be. There may be, for aught we know, a whole world of subconsciousness in the region of reflex activity. The ego's knowledge of itself, or consciousness proper, comes late, and at no distinguishable date. The line between elementary consciousness and self-realization is altogether shadowy, and logically cannot be said to exist at all, since the knowledge of a beginning implies a knowledge of that which is before the beginning. A limit cannot be conceived as existing from one side alone. A thing cannot be known for what it is, until it is known for what it is not; for otherwise it must expand itself without limit in every possible direction, and in every possible quality or state; and so, becoming limitless even in the bare fact of its existence, would be as though it were not.

This opens up to us the consideration of that vast region of vital activity, called automatic. We cannot thrust it aside, and we cannot explain it by the principle of pure mechanism. It has in it, *ex hypothesi*, that factor of the Universe which is not material, or, to be safe, which removes living matter worlds away from dead matter, — a fact freely admitted, as we have seen, by the leaders of scientific thought. In the lowest vitalized form, — in the protoplasmic unit, the life principle is what gives rise to structure or mechanism, and without it no such thing could come to pass.

There seems to be no good reason why the metaphysician and the psychologist should hesitate to go down fearlessly into this world of basic personality and claim for it the heritage which the physicists so freely proffer. If the evolutionist can build up the marvellous human mechanism out of the minimum of structure, and even it confessedly dependent upon vitality as a necessary condition, why may not the psychologist build up the conscious personality out of the positive and antecedent fact which the physicist has to borrow for his structural advance? Surely there is an advantage in the start, and in the end the jump from the highest dumb creature to speech-using man is not more desperate than the leap which the physicist is compelled to make from the highest brute mechanism to that of the lowest human organism, *plus* his psychic nature.

And so with the theologian; if he is not to hold to immediateism in the creation of man, why may he not reverence and adore the Creator in the up-building of, and preparation for, the human personality through innumerable sub-conscious beings of which He himself is the Author and Sustainer, as fully as he can in a discontinuous and orderless creation. For, admitting any sort of orderly sequence and dependence — and who does not? — it is too late to object to intermediations, unless, indeed, the degree and complexity of such dependent sequences is the ground of complaint, — a position few would care to assume. Besides, the theologian is irrevocably committed in the Christian Faith to a stupendous Messianic evolution; and perhaps, when rightly understood, the Personality of the Christ sweeps through and embraces all finite evolutions. But a theory which shall be large enough to embrace Him, who 'lifted with His pierced hands empires off their hinges, and turned the stream of centuries out of its channel, and still governs the ages,' cannot be a one-sided half-truth. When the evolutionist shall add to the theory as commonly propounded, the fulness of that life-factor, acknowledged, but made so little of in the development of mechanism, which implies an ever-active and intelligent Personality, the physical side will have lost nothing, and the metaphysical will feel itself no longer outraged.

Natural selection is a principle recognized as existing prior to the first possible movement towards the development of structure; and implies an active power (nature) behind it — propulsive and purposive at every stage, from start to finish, in the up-building of tissue. At numberless points new principles are necessarily assumed; as sensation, volition, admiration, sense of beauty, fitness, pugnacity, courtship, love of ornament, novelty, sexual affinity, and morality, — all of which belong to the ideal world, and so find their place on the other side of the chasm which confessedly separates matter, as matter, from the conscious world. It is a question of no consequence as to whether these are added successively as they are needed, *ab extra*, by the power of nature (only another name for the ultimate source of all activity, and, according to Mr. Spencer, only not the Infinite Personality of the Theologians because it may be higher), or are potentially in the life-principle from the beginning. The state of the case is not in the least simplified, whether it be assumed as a perfect continuum without break or pause, or as a succession of stages with intervals between. A continuum is inconceivable (as will appear further on), and no mortal wit can bridge the chasm between 'sweet' and 'red,' or between 'natural selection' and 'heredity.'

So far as experience can testify, the breaks are found all along the line; and discover themselves in the most surprising way in human physiology. For example, what is Weber's Law but an irrefutable witness to a lack of continuity in consciousness in response to unbroken gradation in stimuli? What is every heart-beat, every pulsation of nerve-fibre, but the effect of change and discontinuity? I do not see that there is the choice of a pin's point between the action of Nature (the All-Father), by successive starts and stops, or by an unbroken continuity — either being utterly inconceivable. But however else we may think, the evolutionist cannot be permitted, without protest, to shut his eyes to the life-element, with its psychical potentialities, which lies on the other side the chasm which separates mere mechanism from the thought world.

It would be freely admitted by the evolutionist, no doubt, that the vast region of so-called automatic action in animal tissue is due to the life-principle, which, starting in the lowest protoplasmic unit, becomes more and more marked as we ascend towards man; and that the world of reflex action, called instincts, is due to this principle.

This instinct world is a wonderland indeed, so complex and unerring that it takes on the look of intelligence and design. Dr. Carpenter, speaking of jelly-specks (*rhizopods*) says, Suppose a human mason to be put down by the side of a pile of stones of various shapes and sizes, and to be told to build a dome of these, smooth on both surfaces, without using more than the least possible quantity of a very tenacious but very costly cement in holding the stones together. If he accomplished this well, he would receive credit for great intelligence and skill. Yet this is exactly what these little jelly-specks do on a most minute scale; the 'tests' they construct, when highly magnified, bearing comparison with the most skilful masonry of man. From the same sandy bottom, one species picks up the coarser quartz-grains, cements them together with phosphate of iron secreted from its own substance, and thus constructs a flask-shaped 'test' having a short neck and a single large orifice. Another picks up the finest grains and puts them together with the same cement into perfectly spherical 'tests' of the most extraordinary finish, perforated with numerous small pores, disposed at pretty regular intervals. Another selects the minutest sand-grains and the terminal portions of sponge-spicules, and works them up together, apparently with no cement at all, by the mere 'laying' of the spicules into perfect white spheres, like homœopathic globules, each having a single fissured orifice. And another, which makes a straight, many-chambered 'test' that resembles in form the chambered shell of an Orthoceratite — the conical mouth of each chamber projecting into the cavity of the next, while forming the walls of its chambers of ordinary sand-grains rather loosely held together, shapes the conical mouths of the successive chambers by firmly cementing together grains of ferruginous quartz, which it must have picked from the general mass.

Everybody has some knowledge of the wonderful instinctive action of bees, wasps, ants, and other social insects. They present a sort of parody on humanity in their individual community and governmental polity. The Bee is wonderful enough, but the Ant seems to be entitled to claim a fuller round of the virtues and vices of man-life. The 'queen' has her retinue of servants; the community has its architects, laborers, nurses, foragers, physicians, and soldiers. They are not without pampered aristocrats, who 'lord it over' multitudes of miserable slaves captured in battle and 'sold in the shambles.' They have their milch-cows and beasts of burden. Nor does it all look like machine work. They appear to gather information through scouts, assist each other in emergencies, contrive means of meeting difficulties purposely put in their way by experimenters, consult together, — indeed they exhibit in miniature about all that men do consciously, so much so, that some investigators do not hesitate to declare that in their opinion they are possessed of consciousness.

There are, however, insuperable difficulties in the way of such a conclusion, — the very perfectness of their movements telling against the hypothesis. Dr. Carpenter speaks of a little fish (the *chætodon rostratus*), which shoots out drops of a fluid from its prolonged snout, so as to strike insects that happen to be near the surface of the water, thus causing them to fall in, and be brought within its reach. Now by reason of the refraction of light, as he points out, the real place of the insect in the air is not that at which it appears to the eye in the water, but a little below its apparent place; and to this point the aim must be directed. The difference between the real and apparent place, moreover, is not constant, but varies considerably since the rays are bent at different angles at the surface, in consequence of the difference of slant when the insect is directly above or to one side of the fish.

It is surely a little too much to assume that the fish understands the laws of refraction. And besides, if the movements of the lower animal world are the results of conscious intellection then these lower orders are infinitely more intelligent than man, for he does not know how he performs any bodily action, — he is not conscious of what nerves or muscles he uses in any act of locomotion or speech, — indeed, so far as consciousness goes he does not know that he has either nerves or muscles.

But this perfectness of automatic action sometimes leads to very absurd results. Dr. Carpenter gives the experience of Mr. Broderip with a beaver taken very young and kept in his house. Its building instincts showed themselves before it was half grown, when let out of its cage, and materials were put in its way. It would drag a sweeping-brush or warming-pan, taking the long materials first, to the place determined upon for his structure, and placing some of these perpendicularly to the wall, would fill up the area with "hand-brushes, rush-baskets, boots, books, sticks, cloths, dried turf, or anything portable. As the work grew high, he supported himself with his tail, which propped him up admirably; and he would often, after laying on one of his materials, sit up over against it, appearing to consider his work, or, as the country people say, 'judge it'; this pause was sometimes followed by changing the position of the material 'judged,' and sometimes it was left in its place. After he had piled up his materials in one part of the room (for he generally chose the same place), he proceeded to wall up the space between the feet of a chest of drawers which stood at a little distance from it, high enough on its legs to make the bottom a roof for him, using for this purpose dried turf and sticks, which he laid very even, and filling up the interstices with bits of coal, hay, cloth, or anything he could pick up. This last place he seemed to appropriate for his dwelling; the former work seemed to be intended for a dam." As Dr. Carpenter says, nothing could be more absurd from the reasoning point of view, than the attempt of the animal to construct a dam where there was no water, and a house where he was already comfortably lodged.

When we come to the domestic animals we are constantly misled into attributing conscious deliberation and thought to what no doubt belongs to the domain of reflex phenomena. Wonderful as the performances of these creatures are, it is hardly probable that any of them ever reach the stage of self-consciousness. As we have seen in the insect world, the teachings of physiology would have to be all reversed if we were to conclude that, with their lower development of the nervous organism, they have an intellectual eminence which is wanting in far higher orders of development, even in man. It can hardly be contended that even the child has any clearly differentiated self in the first weeks or months of its post-natal existence, — and that after it shows perfectly well marked evidences of thought, such as is worlds beyond what the highest brute ever exhibits; and if this be true, it is not unreasonable to think that no brute ever attains self-knowledge.

Again, it is to be noted that in the animal kingdom there is a fairly well marked inverse ratio in the order of intelligence and instinct. I am aware that this has been disputed; but the arguments against it do not seem to be sufficient to overset the opinion of Cuvier, who asserts it. At any rate, man, while standing at the head of all earthly beings in rational powers, certainly stands at the bottom in his exhibition of that sort of pseudo-intelligence which the creatures below him possess at first hand in such high degree, exhibiting in the ant a complexity and perfection which startles one with its likeness to the slowly and painfully acquired knowledge in man. The calf, the colt, the pig, come into the world thoroughly furnished with the power of muscular co-ordination; while man has to learn every step of the way. They know — (observe here in the use of the word 'know,' how deeply the purely intellectual element is read into reflex action) — they know where their sustenance is to be found, and how to get at it. The infant would perish if it were not helped to the breast; but once there, the one conspicuous automatic function left man out of the infinite store below comes into play; but even that disappears, or is inhibited at will after it has served its purpose; that is, as soon as the child has learned the complicated movements in the business of mastication, and is furnished with the necessary instruments with which to begin operation on solid food. We may well surmise that even this sucking instinct would have disappeared with the numberless others, if it had not been absolutely necessary for the child in its utterly helpless stage. Without this link with the pseudo-intelligent world, all the mothers and nurses on earth could not get an infant over the first months of its existence.

The difference in the lower animals is immense. Nothing is more curious than to see, say, a pig, the moment after birth, trot around, and fight its way in among its fellows for its share of the lacteal supply. Thus there seems to be an order of *devolution*, as well as an order of *evolution*, along which Nature works, — parallel, though bearing an inverse ratio to each other: the one moving from inchoate structure towards greater and greater specialization and complexity; the other, starting in purely instinctive self-motion, loses its blind cleverness by stages, until it all but disappears in man. The one starts in potentiality, out of which more and more complex structural forms emerge; the other starts in determinism, and emerges in potentiality. It seems probable that this inverse ratio could be traced backward along the whole line of the evolutionists, but I can only illustrate what I mean. Thus, though man has been a speech-using animal all through the ages, no child ever comes into the world with a word ready formed on his lips, while the potentiality or capacity for language and the mechanism necessary therefore has been greatly increased. The Flycatcher, just out of the shell, strikes at and captures an insect with the utmost precision; the infant, with a native power to master the problems of the stellar depths, knows nothing at the start of directions and distances. This, it seems to me, is a fundamental distinction. Mechanism is potentiality fettered by determinism; personality is determinism swallowed up in potentiality.

It may not be out of place to say that in all this I am far from implying anything touching the ultimate destiny of the inferior animals. I have no desire to go beyond Bishop Butler, who declares any discrimination against the brute creation in this regard to be both 'invidious and weak,' speaking of it with scorn as 'the invidious thing.' Nor do I ignore the fact that some, perhaps all animals, are teachable, and have understanding in some degree, and so are conscious, though not self-conscious; while, on the other hand, man is, in the beginning, and remains through life, dependent in great measure upon the reflex mechanism of his nature, yet he is almost without instinct proper, that is, action which has the look of intelligent forecasting, but is really the result of a blind propulsion. Instinct is gradually replaced in the ascending order of nature by understanding; and whatever can be learned by conscious effort is left to be so acquired: and thus it is that man, infinite in power and wisdom potentially, is at birth the most helpless and ignorant of all created beings.

CHAPTER XI.

THE CONCEPT-FORMING PROCESS.

Muscular co-ordination. Education of the organism. Vital organs not under control of the will. Analogous psychical conditions. Process of thought development. Like and unlike. Discovery of meaning. Attention. Retention. Concepts. Concept-masses. Apperception. Thought as thought. Language. Introspection. 'Pure' and 'Empirical Ego.' The 'One' and the 'Many.' A sense of knowing deeper than 'understanding.' Personality antedates knowledge.

THE human organism in its infinite complexity is complete at birth. The muscles are already equal to considerable effort. It is not for lack of strength that the infant does not use hands and feet at once, but because it does not know how to co-ordinate the complex system of muscles. This co-ordinating power has to be gradually acquired through conscious effort. Even the reflex-centres, in large part, have to learn their work. This education is accomplished in the beginning through a low order of consciousness,—sub-consciousness, indeed. The self does not know what the office of any organ of the body is, nor where it is, nor even that there are any organs; but the vital tides, so to speak, surge to and fro throughout the mechanism, and the infant finds, little by little, that it has power to control the movements of the body, and purposive effort follows. The exercise of distinctive purposive action doubtless long precedes any proper recognition of the end in view, or the means by which it is reached.

There is a constant play between the three fundamental modes of the personality — sensibility, cognition, and conation — each performing its office because it cannot help it; while the reflex nerve-centres learn their parts in response to volitional command, at first painfully and slowly, until at last they perform their functions with automatic promptness. In learning to play on an instrument, to sew, or to knit, everybody knows how slowly and awkwardly the fingers respond to the bidding of the will; and how many times a movement must be gone over before smoothness and certainty can be had; but after awhile, the fingers seem to do the work of themselves. What is more striking, they seem to be able to do their parts better, in mere mechanical actions, without thought than with it. Take one who has a little knowledge of (say) the flute, let him work out a few tunes from the notes with the tyro flutist's usual persistence, until they are well memorized. He will now play them better when unconscious of effort than when he bends his whole soul to the performance. Moreover, the fingers will now do for him in their office, what he for his life cannot tell them in advance they ought to do. If you ask our supposed youth, for example, what the fingering is in the fifth bar of any familiar piece, or what notes will be used in transposing that bar to another familiar key, he cannot at once tell you; but the fingers will make their dispositions in either case without the slightest hesitancy as they come to them in order. In this case there has been conscious effort at every step along the way; but the reflex centres when they have once thoroughly learned their part are rather hindered than helped by any officious interference on the part of the understanding. This work of translating thought and purpose into automatic action goes on from the earliest stages of infancy till the close of life, so that it may be said at last that we belong to our habits, rather than that our habits belong to us. The simple truth is, man is born a co-ordinator and unifier; and he begins his work, by virtue of a propelling power which comes with him somehow, long before he knows what being in the world means; and continues long after he has found out, or has given it up. This power is a simple fact, wrought into his nature by the All-Father, or, if one likes it better, by that 'Unknown Cause of things' which, if not personal, is 'something higher than Personality.' Through this power, which man can neither get out of his nature nor explain, the fruits of conscious work are all the time passing into the unconscious, and, no doubt, essential personality, in its two-fold factors — physical and psychical.

But let us be careful to distinguish at this point. As has been often said already, the human organism is full of reflex activities, but of instincts proper, that is, of activities which appear to be intelligently directed but are not, it has a very minimum. The mere mechanism upon which the psychical factor of the personality is superposed is in a great degree independent of conscious energy. Most of the organs of the body do their work without thought, and in spite of it. The heart with the whole vascular system, the respiratory organs, the sympathetic and gastric systems, and even the nervous system, in large part, are essentially automatic, out of and beyond the control of volition; and this for the plain reason that they could not be trusted to volitional control. If the heart and lungs were dependent upon the will, we could not maintain our existence. We should die in some moment of forgetfulness, or in sleep. The mechanism of the personality is precontrived for us, and regulates itself. We are given a limited control, indeed, but only by violence can we stop or resist its action. But in all this, it will be observed, no knowledge or semblance of knowledge comes to us. The organism will not, untaught, draw a straight line for us, nor shape a letter, utter a word, or tell us a single fact in mathematics, geography, or physics. It only stands ready to help us to all these, and to untold stores of knowledge beyond. Thus man *potentially* knows all that can be known, but *actually* knows nothing whatever.

In like way, there are certain fixed and necessary elements discoverable in the psychic factor. We must think, and feel, and act; we must combine and contrive; and in exerting these psychical activities, we fall upon certain absolutely necessary truths which we neither make, nor are able to unmake; as that we exist, and things exist, — that one thing is not another thing, — and that three things are more than two things, and all that: but not one of these is a ready formed notion in the mind. In between, there is a vast area of purely volitional control, and through it we are constantly changing, for the better or worse, both the physical and psychical factors.

It is no good to resist or complain: we are given a certain capital; we must use and increase it, or abuse and lose it. To express it in another way: nature gives man *power* to know, but furnishes him with no ready-made knowledge.

Through external stimuli, and by means of the several organs of sense, the rational nerve-centres are forced to respond in certain reactions, which would remain simply and forever reactions and nothing more, if it were not for the psychical power of co-ordinating and arranging. No single sensory effect could have meaning, and no thousands, so long as each one remained isolated and unlinked to the others. It is not enough that they shall simply *be*, or that they shall be, in themselves, like or unlike, — precede and follow, — but the self must discover this before the notion of agreement and difference can arise in the personality. It is not until after these successive and varying reactions have come before the co-ordinating and unifying activity of the self, over and over again, that the new thing, 'meaning,' stands out in consciousness. If a rose be the stimulus, the vibrations of the luminiferous ether are not 'red,' the irritations of the olfactory nerve are not 'sweet,' nor are the tremors of the muscles of the fingers 'heavy,' nor are they all together in any wise *like* the notion of the rose in the mind. They are each and all of them but the physical signs by which the self reads into them the meaning which is the mental rose.

This may be understood by a telegraphic message. The ticking of the instrument might go on forever, without any message, if there were no operator who knew the signification of the clicks; but by one who understands what to most of us are but senseless sounds, they are woven together until the meaning is clear. This power to 'understand' is so familiar to us that we can hardly enter into its wonder. 'Meaning' is wholly unlike, in degree and kind, the electrical impulses along the wire, or the ticking which results.

The electrical transmission is analogous to the action of the sensory nerve; the click of the instrument may be compared to the reaction of the nerve-cell, but the meaning is only in the mind of the operator. Motion in any form is but a fact of mechanism, and its physical results are but dead symbols; in themselves wholly meaningless until the psychical factor appears upon the scene. Between this factor and the symbols is a world-wide chasm over which no conceivable bridge has ever been built by the wit of man; and yet this chasm is crossed and recrossed in fact, every moment of our lives.

We see, then, that it is not any individual nerve-action, nor any succession of them that is sensation, but it is their interpretation through the arranging and co-ordinating power of the self. Sensation proper is thus not physical, though it has a physical basis, but belongs to the psychical side of personality. Now, in order that there shall be arrangement, there must be attention or conation; so that the action of the inchoate will must be present in interpreting the most elementary signs which come to the self from the external world. But, as we have seen, attention to individual or isolated sensory actions would accomplish nothing, — there would be no meaning in a bare fact out of relation; so that there is another element necessary before any co-ordination shall take place; it is retention. The present sensory act must be put alongside of, and compared with, other such acts, and the likeness or difference noted. But all save the present act are gone: they must therefore be brought back by some power; and this power we recognize in retention, or the rudimentary memory. We have now all that is necessary to the formation of a concept, — divers sensory acts held together by the elementary memory, attention by which activity is directed, and the understanding by which they are seen to agree or disagree. A 'concept' (*con-capio*, to bring together) is thus the product or joint action of all the several modes of the ego, sensation, cognition, and conation, and, if I may so say, on the field of memory and in the light of consciousness.

In the above elementary exposition of the concept-forming process, we have left out of sight all details so as to present the matter simply and connectedly. It has, perhaps, been sufficiently indicated that there are many stages between the dawn of conscious recognition of things and events, and the full day of external perception and self-introspection.

In all the earlier stages, concepts are vague and ill-formed — they have been called gathering mists or cloud-masses. These are ever changing, through the rushing in of newly formed concepts, and, as in the beginning the child has no prejudices (which is only another way of saying that its concepts have not yet become permanent), these masses are ever breaking up and reforming under the action of accumulating material. Gradually those notions which have gathered into themselves as a centre the multiform fragments of half-formed concepts, endure and become for us the real and actual.

This reaction of mind in unifying the materials of knowledge is called 'apperception.' The older and better formed concepts (to continue our material simile), meet and absorb, or modify the newer. The older concept is called the 'apperceiving'— the newer, the 'apperceived.' The once formed and measurably permanent notion has a great advantage over one just emerging in consciousness. It will attract to itself what is like, and reject what is incompatible in the new. There is thus a fairly steady growth, and an increasing confidence in our own ideally-real world; and this becomes stronger until no sort of new notion can gain a permanent footing in the presentative, or sense-perception area; as well as in another region — the rational or *a priori* area, of which we shall have a good deal to say further on.

But there is still another and a less permanent field — what may be called the region of contingent and speculative knowledge, in which radical changes may be wrought by the presentation of new matter. Older apperceptive strongholds are sometimes broken up, and the ordinary process reversed; but this becomes more and more rare as age advances.

It will be seen that the simplest concept is a fasciculus or bundle, never one and single. We have here once more, the 'many in one.' In its formation we have the process of differentiation, or differencing, inasmuch as the individual elements which compose the complex whole must be distinguished, before they can be united. We have also the process of 'integration,' or bringing together into unity. That is to say, we have rudimentary analysis and synthesis; and each implies the other, inasmuch as there can be no whole, without a recognition of parts; and no parts, without the recognition of the whole. They are, thus, but the two phases of the same act, like the fabled shield which was both 'argent and gold.'

It is plain, also, that there must be discrimination or comparison in the lowest concept formation. Thus the thought-power, in its elementary form, reaches down to the very bottom of all sense-perception; but in this low form it does not stand out as clearly differentiated thought. It is not until the process of apperception is well advanced that we recognize thought as thought. When concept is compared with concept, the purposive action of the will becomes manifest, and we have what has been called 'redintegration.' We may now be properly said to think; and the product of the thought is a larger or resulting concept called a judgment. Thus, a judgment may be defined to be the recognition of the agreement or disagreement of two concepts.

But the two concepts so compared must have each a mental form or sign, else they could not be recognized so as to be compared. These signs are subjective words, and when they are given articulate, or any sort of expression, they are language. Taken with this modification, Professor Max Müller, and those who preceded him, are certainly right in the declaration that there can be no thought without language. We shall have to return to this further on.

Our knowledge of the outer world is built up, as we have seen, through sense-perception; but there is a vast domain which the senses do not discover, but grows out of those psychical limitations spoken of a moment ago. Such notions as 'truth,' and 'right,' and 'mercy' have no discoverable element in common with sense-perception. One cannot see 'mercy' with the eye, nor discover it by any other sense; and it has no likeness to anything in external nature. There must be then another and a higher domain than the external, and the way to it is quite different from that through the senses. What we have been so far speaking of may be called 'sensuous truth'; this higher region 'rational truth.' I only mention it here, that we may not hastily conclude that the world of sense furnishes us with all knowledge.

Again, there is self-knowledge, or a comprehension of the fact that we 'know,' and what we know,—the knowledge of our own thoughts and emotions,—the phenomenon of self-consciousness. The knowledge of the body belongs, of course, to sense-perceptions. My hand is an external object to my mind, as much so as a book or table: so is any member, organ, or tissue, or so much of any one as can be made an object of sense-perception.

In like manner, I not only think and feel, but I can make my thought or feeling an object of thought. Thus I turn my thought-power inward, and think upon the notions or states of the self. Now these notions or states are not the self, any more than the hand or the foot is the self. They are, however, phenomena of the self. Everything I know of, or in, the self, is an experience of the ego; and all these put together constitute what is called the 'empirical ego'; while the subject of all such experiences — the essential self — is called the 'pure ego.'

Now it is the 'pure ego' that is the one, original and indisputable fact for every man; and so, known to each of us in a way no other fact can be known. It is the basic truth to which every other truth must be referred — the centre and source of all certainty. As such it cannot be tried or measured by any other truth whatever. If it were so, it would at once follow that there could be a higher and more certain truth. It is therefore ultimate, self-asserting, single. It is not possible to compare it with anything else, because it is a *unicum*, without place or parts, and cannot be said to be known in thought at all, since thought implies relations. I know it all the time and every how — in sensation, thought and will; but these phenomena do not prove it to me. It is susceptible of no proof, for it is beforehand with all possible proof. It is not susceptible of definition, for its sole limitation is that which involves the self as a necessary factor. To say, 'I am I' is an identical proposition which advances nothing. To say, 'I am not, not I,' separates it indeed from all but itself, but that not-itself depends solely upon the self for its content.

And yet the self exhibits modes and activities; feeling, thought, and action are its notes, or marks; and they have no meaning apart from it. Here we are brought face to face again with the eternal riddle of philosophy — the 'one and the many.' As we have already seen, we can neither understand the 'one,' nor the 'many' except as either acts as the background of the other. The one cannot have parts, nor sides, nor phases — since it would then cease to be 'one.' It cannot have place in space nor in time: it must be out of all possible relation; and so have nothing left by which to make itself known; and thus the manifold must be brought back before the one can have any sort of meaning for us.

But the 'many' is in just as bad case. It can have no meaning without the one, for each attribute, sign, or note must be recognized as 'one' before it can be cognized at all. The many cannot be many except it be composed of elements or units. But all this lies in the province of metaphysic, and is familiar to every student of philosophy. It is as old as Parmenides, and found everywhere through Plato. It is hydra-headed, always starting up whenever we force our thought into the region of the ultimate.

Now, we do without doubt know the 'one,' and no plain man ever suspects that there is any trouble about it, and perhaps would fail to see any after the most persistent efforts to point it out; and so with the 'many.' There is a sense of knowing, therefore, which is deeper than that of mere intellection. What we commonly call the understanding must deal with subject and predicate. That it is at all, depends upon the shackles in which it is bound. It cannot transcend the trammels of limitation; but the self, as we have seen, is not the mere intellect, — the intellect being itself but a mode of the self, and dependent upon it. The self is, therefore, logically prior to the first dawn of the intellect and is its source and support. It comes, somehow, at first hand from the source of all giving.

There is a sense, I repeat, in which we use the word 'know,' that is deeper than that in which we use it when we speak of a cognition through the understanding. It is the sense in which I know my own existence, though I cannot construe or explain it to myself or another — the sense in which I know my thought to be mine, and one thought to be different from another. It is the ground of all first-hand knowledge, the only warrant, ultimate truth of any kind can have. It comes nearest to being pure feeling, perhaps, and is a condition of all knowing in the sphere of relations. Truths of this primordial character, I can reflect upon, note their activities, and wonder at; but cannot construe — they have no *quale* or whatness for me. Thus, while I know that I am what I am in my essential personality, by a necessity that is absolute and inexplicable, the self which I construe, I know only through my modes and activities. This empirical knowledge has the self for its subject, and is what we call introspection, — self-consciousness. It is the manifestation of the self to the self — the intellectual apprehension of the 'empirical ego.'

This introspection or self-conscious knowledge is difficult, and therefore late in development. It undoubtedly lags considerably behind sense-perceptions, and like it undergoes the apperceiving process. It is thus shifting and variable, growing into greater and greater stability. This is why people know themselves so little, and so different from what they appear to be as seen by others — felicitously put by Dr. Holmes in his "Three Johns." Out of this mass of self-knowledge, there is built up a concept, appearing no one knows exactly when, which we call the 'me' — the 'myself.' This also belongs to the intellectual zone of our being, and is the result of the some time continued action of the 'pure ego.' This 'pure ego' stands, from an epistemological point of view, in quite an analogous category with the substance or 'itness' of any material object, with the difference that the substance of the object gets its meaning wholly from the ego. Personality, which is but another name for the 'pure ego,' antedates all understanding of itself, and it is in the active enjoyment of its manifold prerogatives long before it has any explicit knowledge of its heritage, much as an infant king is actually a sovereign long before he has any apprehension of the dignity to which he is born.

CHAPTER XII.

THE FUNDAMENTAL MODES OF THE SELF. — MEMORY.

Consciousness. Differentiation of Feeling. Of Cognition. Of Will. An end ideally first. Self-development. Perception. Intuition. Ideas in the mind not *like* objects without. Space. Time. Memory. Mechanical basis. Objection of Lotze. Complexity. Illustration from Sound. Phenomena explicable upon theory of mechanical basis. Dr. Rush's case. Dr. Carpenter's Welshman. Coleridge's case. Power to recall the past. Sudden recollections. Law of Association.

CONSCIOUSNESS may be called the daylight of the self. Whenever any psychical effort or reaction rises above the intellectual horizon, — which may be called the threshold of cognition, — the self recognizes it, and this recognition is what we call understanding. But the line of separation between the conscious and the unconscious events of Personality, is not definite and exact, but vague and shadowy, like the separation between night and day. A vast amount of the life-history of the ego lies below the illuminated circle, in the totally obscure or dimly-illuminated region. All reflex action proper, and all the rudimentary work of apperception lie either below the horizon of consciousness, or in the border-land. This is not only true in the initial stages of existence, and during the gradual evolution of self-recognition, but it is true all through life to its very end. Thus it is that we know nothing of the greater part of the bodily movements under our control; nor of much that goes on about us at any time; while we know nothing of what takes place when we are asleep, or otherwise unconscious. It would be intolerable if one had to be conscious of the movement of every muscle, the action of every nerve-fibre, and of every object and event about one in the external world. It should seem that one of the chief functions of the 'inhibitory' mechanism is to produce silence, so to speak, in the auditorium of consciousness, when attention is not necessary, somewhat as the musician lays his hands upon the strings of the harp to stop their vibration when they have done their office.

Now, as a fact, out of the sub-conscious personality, either by the development of a new and higher physical basis of psychical action, or from the further and more delicate differentiation of existing nerve organisms, there does emerge what we have called the dawn and daylight of the self. It is only then that we begin to recognize the three fundamental conscious activities of the self, so repeatedly spoken of, — sensation, cognition, and conation.

Let us look at these separately for a moment. First, sensation, which shows itself from the protozoa up, has been an active agent in the work of tissue building in all reflex mechanisms on to the evolution of the psychical centres; but when consciousness, in its own right, appears upon the scene, it takes an immense flight, and becomes what we call 'feeling.' With it comes a new heaven and a new earth. Old things have passed away. It is a translation into light out of darkness. It is true, the personality cannot maintain itself wholly in the new regions so revealed. The roots must still remain in the soil below, and so the immense mechanical work which is necessary as a substratum of this new world of feeling is carried on still by the same sub-conscious agencies.

With regard to cognition, or the power which discovers meaning in things and events, that, too, has been working all through the processes of advancement. It is, indeed, difficult to form any notion of what 'meaning' can be when not lighted up by consciousness; but, when rightly considered, not more difficult than to conceive what unconstrued sensation can be. We, from the sphere of consciousness, are compelled to read into it, as well as into sub-conscious sensation, what conscious experience has taught us. There is, however, a solution of this which must be admitted as a possible hypothesis, and which to many seems the true way of regarding it. It is that the personality through the whole of its sub-conscious period has its feeling and its thought in and through the Infinite Personality, — that as the Ultimate Causative Power is leading, or propelling all processes on to an end, He supplies everything that is needful; so that sensation and thought find, all along through the darkness, their reality and light in Him, much as the lower animal world is helped forward by instinct.

The case is quite analogous with the development of conative or elementary effort into purposive Will. It seems plain that the whole animal creation below man cannot be set down as mere machines, without including man in the same purely mechanical category; but as we have seen, there really seems to be no school of thought which holds such a radical theory. The lower order of animals not being then mere machines, conscious action must be attributed to some of them; and if to some, then, to those below unless a clear line of separation can be found. But this, it is confessed, is impossible. Where, then, are we to stop? Effort of some sort, active or passive, must be admitted down to the protoplasmic unit, and there seems to be no halting place until we reach it.

At the point of its emergence into the light of consciousness, however, we recognize it first in what we call attention. It then begins to stand out more and more distinctly and in fuller proportions until at last we discover the fact that the sovereign principle of the personality — the will — has quietly seized the reins and conscious self-determination has begun.

Not until now, and perhaps not for some time after, does the concept of the self, clear and certain, stand out to he self;

so true is it that that which is first in idea and essence, and for the sake of which all intermediate stages have value, is last in manifestation. The seed is for the tree, the tree for the bloom, the bloom for the fruit. The end of this sequence was at the beginning in the seed, unless we are to give up all meaning whatever. It is the same in the works of man. Take a single illustration. Workmen break up and drag forth great masses of stone from the quarry; others toil at transporting them to the sea-shore; again they are seized and carried forth,— thrown overboard and sink out of sight. What folly to one not in the secret! But gradually an artificial island rises above the surface; and on this a tall shaft shoots up, and finally a blaze of light streams far over the 'waste of waters,' a beacon of safety to the mariner, while warning him of danger. The engineer who planned all this, and carried it on through its successive stages, saw that blaze of light from the beginning; and only in the reality of that first light in the mind, is there reality in the final blaze on the sea. And so with anything that has meaning. And thus it is that the ideal and the real are both beginning and end.

It is not until after this reality — the self, both new and old — has, so to speak, attained its majority, that it enters upon its distinctively purposive existence. A new factor has appeared upon the stage as marvellous as consciousness itself, the power of self-direction and self-limitation; and we find ourselves masters of our own movements within large limits, mechanical and psychical, imposed by the Author of existence.

We now arrive at the true self-formative stage of personal development. Up to this time, the personality has been chiefly subject to the push and pull of the world-evolver, and the work accomplished has been the developing of organs, and their functions, for the true work to be done with them by the self in its own up-building. While the mighty power behind all phenomena does not cease to sustain and propel, now that It has vested the self with this highest prerogative, self-direction, It confines its offices to those of the silent and humble sort, devolving responsibility on the self in the ratio of its elevation. The self enters upon its newly discovered functions with full authority of self-direction, and a large province in which it has the power to carry its purposes into effect. Its work is, in legal parlance, to 'reduce the Universe to possession,' building up within a self-world with greater or less fidelity to the reality without and above it; refining, strengthening, and enlarging itself in the three-fold modes of thought, feeling, and conduct. The world known, is the objectified self; the self knowing, is the subjectified world. The self can know nothing of reality which has not in thought entered into and become in so far itself. A landscape lies spread out; it may be miles away: I know it by the reactions in my mechanism, due to sensory changes. On the other hand, whenever I try to think upon myself, it is some phase or other of the empirical ego which engages my attention, and so is in so far an element of the non-ego.

If the work of the self upon the material furnished it in the reality without be well and faithfully done, the self-world will be true; if falsely and faithlessly done it will be false, and the personality, living thus in a false environment, must be itself false in one or other, — it may be in all its modes. This we shall understand better farther on.

This process of seizing upon, and bringing in the Universe, is progressive and complicated. It is partly due to environment, and partly the result of purposive and self-directed action; but the self in its highest mode is guarded by its nature against whatever in environment has not its active co-operation. Perception, or it may be called 'thing-knowledge,' is forced upon the self through external stimuli, and is, in so far, mechanical; but whatever use the self makes of such crude material will have value from the character of the purposive element which enters the action.

Perception, in its proper psychological sense, is the immediate knowledge of 'thing' now, and here present as a stimulus to psychical action,—'thing' being taken in its largest sense. 'Perception' is often used loosely, to embrace any notion in the mind, whatever be the content; as when one says one perceives (comprehends) the truth of a mathematical proposition; perceives (understands) the excellence of a method, or the advantage of a particular course of action. Perception in the stricter sense must have an objective reality as its content. An event may be the object of a perception, but in strictness it is not the event itself, but the mechanical element is the necessary factor, and it is for this reason that it is so largely forced upon the self by environment. The perceived-world is just as much of actuality as is present to consciousness, through the action of stimuli, at the moment of such action.

It is not our object to discuss problems, and therefore it is not necessary to enter upon the many phases presented by different theories of perception. Whether an object is seen immediately, *i.e.* seen face to face, and in its own right; or, mediately, *i.e.* through the agency of something not the object perceived, has engaged the earnest inquiry of philosophers in the past. In our view of the case, it is either, or both, as the inquiry is presented. It is to be remembered that the body is not the self, nor are the nerves, and nerve-centres, nor the whole of the personal mechanism, what I know in consciousness as 'me.' The self is not shut up in the brain, nor is there any power to find, or to conceive of where it is, or what it is. With this carefully in mind it seems a rather barren question to ask whether the self knows any particular object, as a tree, or is only told about it by something else. The only knowing that is done in the case, is done by the self, for certainly neither the nerves, nor the air, nor the luminiferous ether know anything about it; and can tell nothing. Again, what is the tree to be known? Of course the phenomenal tree. Nobody contends that we can know the de-phenomenalized or essential tree. Now where are phenomena known? at the end-organs of the fingers, or the retina of the eye, — or at the tip of the tongue? Surely not; nor yet at the nerve-centres. Knowledge is in, and through, all these, however, and whenever they may be stimulated; and all knowledge is in the self, and not at the tree nor any*where*. It has no space form; and so the question as to representation or immediate perception has really no substantial meaning. All we know, or ever can know, about an object is a concept of the object of greater or less fidelity to the externally subsisting truth; and just as this knowledge becomes fuller and more accurate, the self has reduced more and more truth of the Universe to possession.

When it is contended that we see the very tree itself, we say, indeed we do. We see just that which the tree was designed to show forth in color, and in every other way of appeal to us. We have no thought of any agencies or intermediations. We are directly conscious of the only sort of tree there is for us, — that of which we have a notion in the mind, and project as a necessary part of that notion in space, just at that place, and in just so much of it, as is in that notion. If it be said, But we may be mistaken; and may find that the actual tree, and our concept of the tree, do not agree, such an objection arises from the failure to take in rightly what the relation between the concept and the actual tree really is. The concept is not necessarily, perhaps never is perfectly true to the object; but it is at any moment exactly what we see and know of the object at that moment. It may be mended a moment after, and then we shall have a juster look at the tree; but this in no wise affects the reality of our first look. It would be quite as reasonable for one looking at an object through an ill-adjusted telescope to say he was not looking at the object at all, because the next minute, by better adjustment, he could see the object more clearly.

It may be well to caution the reader against the notion that the concept of an object is in any manner *like* the object— like in a pictorial sense. If one could look into the mind of another with every possible degree of refinement in detail, we should see no set of little pictures or images; nor anything more resembling the objects of sense without, than the written characters which tell me of my friend's good fortune are like my friend, or the event. These are in the mind signs, or whatever else they may be called, which the self can spread out into a picture, or a truth, but likeness there certainly is none.

Now, how it is that the concept of externality or 'outness' arises in the mind in the beginning does not much concern us to inquire, except perhaps as a curious question. It has been much discussed, and there are some difficulties about it in a philosophical sense; but none whatever in a practical way. One pushes out the hand originally because, perhaps, one is alive. It does not seem that any great difficulty should be felt that a living organism moves. This means change of place in whole or part, and room for movement; so that there is no lack of material out of which to build up the concept of space. Whether motion alone with the sense of touch would give rise to any right notion of form and figure has been vigorously disputed; but with the sense of sight superadded, the practical result is clear; and that is sufficient for our purposes.

Space is thus a necessary element of perception. Since a perception must be of an object or event now present, Time enters as a necessary factor. We do not now open the inquiry as to what time is; but in its empirical aspects it presents difficulties enough. Like space, there are no practical difficulties, for we know all about Time, as St. Augustine has said long ago, if we do not attempt to explain. But when we say '*now* present,' it may reasonably be asked what we mean by now; and the answer is not quite obvious. We may say that we mean to exclude the past, and not to include the future. But what is left when these are gone? The past claims the second or the least portion of the second which was just here, and the future will not lend the present the most infinitesimal part of a second which has not yet arrived; so that the present is reduced to zero, or at most to a mathematical line without breadth, separating the past from the future. The past moment has vanished, the present is without duration, and the future certainly is not yet. But for all that the present is a very real somewhat, — perhaps in strictness the only time-reality. Indeed, whatever is above the threshold of cognition is now present, and continues to be the present as long as it continues within the illuminated circle of consciousness. The change which takes place in this field of view is what we call succession. Time can hardly be said to have length, except in a borrowed sense, since length is a dimension, and belongs to space, and time and space are incommensurate.

But do we not know the past? Certainly; but whatever we know of it is present. All the past there is for us, — that is to say, all that we do know, or ever shall know, of what was once within the sphere of consciousness must come again within that sphere, and be present once more, before it can be known. The past is the present knowledge of psychical conditions already experienced; the future is the present knowledge of psychical conditions in anticipation of experiences to come.

That power or instrumentality which is charged with the office of bringing past acts of consciousness back once more to the present is called memory. We have seen it all along in a humble sphere, where it was called ' retention.' It is no new thing, but has now passed from its inchoate form, and entered upon its time office within the illuminated circle of consciousness. It had to be, in some rudimentary form in the earliest manifestations of animated nature, existing far below the line of consciousness, — perhaps even in the elementary protoplasmic mass, — in order that any sort of unity might obtain among the varying phases of the several life-modes. Retention, thus, is, so far as we know, simple and fundamental, while memory is complex and highly specialized, retaining its original biological character, but with a psychic factor superadded which quite transforms and sublimates it. In its rudimentary form its office is to retain or link together the states and activities of the living organism; in its psychic form to preserve the continuity of the acts of consciousness. Thus it does not simply conserve the individual facts once present to the mind as so many units. It is the concept-continuum or plexus of the conscious personality. It is the unifying bond of the empirical ego, which is constantly undergoing change, precisely as in the process of apperception. What we know of a thing at any moment — which is only another way of saying, what we are at any moment with respect to the thing — that the memory preserves to us. As this knowledge is made up of many elements superposed and blended into a complex whole, so with memory. Any special deliverance of memory may be compared to the composite effect produced by superposing many faces upon one photographic plate. Each individual leaves its effect behind, but is modified by each in turn. The first glance of any object gives rise to some modification of the psychical mechanism. The next changes or fills out the first, and so on through any number of observations, or through one entire and continuous scrutiny. No two observations — no two consecutive reactions of the psychical cells are precisely the same in perception, but all blend into each other, giving rise to greater distinctness at some points, and less at others; and this change continues indefinitely, though becoming more and more stable by persistence. This marvellous composite effect in perception and in thought is committed to the keeping of memory.

It is thus the nexus of the empirical ego, and gives stability to each modified concept, maintaining the continuity between successive acts of knowledge.

Now, happily memory does not force upon us all at once, and all the time, what it is ready enough to present, or rather *re*-present; nor is it so officious as to retain at all, in any distinct form, much of what was once in consciousness. Life would indeed be intolerable if one could forget nothing. In one sense, there is nothing forgotten, — that is to say, every psychical energy is, doubtless, gathered up in the composite concept; but in the blending process much is forgotten and lost to consciousness.

The physiology of memory has been earnestly discussed, and the reasons presented for holding that it has a mechanical basis in cerebral action are exceedingly strong, — indeed, too strong to admit of successful refutation. The metaphysicians, as a rule, have set themselves against the physiologists in this regard; but, it should seem, the contest is unequal, and the point not worth contending for. No one in these days pretends to deny that the brain-cells play an important part in sense-perception, or the 'presentative' power; and this conceded, it is hopeless, even if there were any point to be saved, to deny them some part also in the '*re*-presentative' power.

The difficulty urged by Lotze, namely, that the reactions of the cerebral centres are infinitely varied in every act of perception, as when "we see some one approaching, every step nearer he comes, the image on our retina assumes larger dimensions; hardly one point of the whole figure answers at any one moment to the same spot of the eye as at the moment before; not one after-image, but numberless images all different one from another would remain, if our nervous organs fixed every momentary impression in permanent traces," seems to tell just as heavily against the power of the self to read at any moment a definite perception into this 'agglomeration' of brain-action — a position he seems to accept without question — as against its being able to do the same thing in the after-effects of these same signs. That motion, however infinitesimal, when communicated to masses, must by the principle of inertia continue until dissipated, or converted into some other form of energy, is too well established in molecular physics to be easily shaken; so that it seems impossible to dispose of the brain-states instantly at least; and if they continue for one instant after the action of the stimulus is withdrawn, why not for the next; and where shall a limit be found? And if they continue at all, there seems no earthly reason that they should not be intelligible, just so long, and in such degree, as they continue to subsist.

The objection on account of their complexity is one that may well stagger us; but science has already made such demands upon our credulity in this regard, that it is now too late to halt upon a mere question of degree. Take, for example, that experiment in sound, given already upon the authority of Dr. Tyndall, in which a piano, two floors below the lecture-room, sends up its vibration by means of a rod resting on the sounding-board. Reflect upon the fact that each of the wires struck has upon it, in addition to its fundamental swing, a multitude of superposed vibrations, or over-tones, — that the sounding-board has to break up into an infinite number of vibrating segments with fixed nodal lines between, in order to respond to each of the strings and their innumerable 'partials,' — that somehow these all, without losing their independent existence, have to find their way to the sharpened end of the rod, and then interlace or pack themselves together without loss or confusion, so as to accommodate themselves to the quite different form of the rod, and then spread out over the resonating body at the upper end, — itself repeating under changed conditions the complexity of the sounding-board, and then on through the air to the still more marvellous mechanism of the ear, — what room is there, in the light of such bewildering complexity, to stand out upon the ground of mechanical difficulties! But all this, we have every reason to think, is but the beginning of the intricacies of motion in nature, when we consider the demand made upon us in molecular physics.

In the phonograph we have what seems to answer in a remarkable way to what is supposed to be the action of the nerve-cells in the case of memory. In it, the amazing fact confronts us that the human voice can be, so to speak, wrought into a waxy composition in a jagged line plowed by the point of a stylus in such wise that the infinitesimal variations of the impressions, upon reversing the process, give back the words with the tone, accent, and timbre imparted to them by the voice! One feels, notwithstanding the evidence of eye and ear, that it must be some unread page of the Arabian Nights' Entertainment, so far is it beyond the powers of the understanding; and yet all this, for delicacy and rapidity of movement, stands untold degrees below the phenomena of light, heat, and electricity; and who shall say how far these in turn are removed from the possible refinement of motion in gravitation, chemical affinity, and vital action!

We cannot stop to bring together in any detail the physiological reasons for holding that the brain, somehow, retains in its molecular structure the signs of past cognitions, but it will be well to mention some of the many phenomena which find explanation upon this hypothesis.

First, injury to, or the removal of certain areas of the brain, as we have seen, cuts off all recollection of objects previously known, as in the case of an animal when food is presented. The power of perception is, of course, involved; but it will be remembered that memory is absolutely necessary to any act of perception. Besides, there are many cases on record of people who have received injuries of a special area of the brain, who, although they continued to understand perfectly what they wanted to say, had lost the recollection of the proper word or words necessary to convey their meaning. The connection between thought and language is very extraordinary. Dr. Abercrombie mentions the case of a man who had lost the knowledge of spoken names, though he knew them well enough if written; and, retaining the sound of such a word in his mind, upon looking in a list containing the word, when his eye fell upon it, would recognize it at once. In cases of the softening of the brain, and in the cerebral changes due to old age, memory is greatly affected; sometimes almost wholly lost.

There are many other phenomena which find their only scientific explanation in impeded or peculiarly excited cerebral action. For example, the well-known case of the student in Philadelphia, reported by Dr. Rush, who, on recovering from a fever, had lost all his acquired knowledge. "When his health was restored, he began to apply himself to the Latin grammar, had passed through the elementary parts, and was beginning to construe, when, one day, in making a strong effort to recollect a part of his lesson, the whole of his lost impressions suddenly returned to his mind"; due doubtless to the removal or absorption of some stoppage in blood circulation.

Dr. Carpenter gives an account of an old Welshman, who, separated from all who spoke Welsh for fifty years, found himself entirely unable to understand his relatives, who, on a visit to him, spoke in their mother-tongue; but in an attack of fever when past seventy, he talked Welsh fluently. Cases not unlike this are quite common.

The most extraordinary case of this sort is given on the authority of Coleridge. In a town in Germany, a young woman, who could neither read nor write, was seized with a fever, and began to talk in Latin, Greek, and Hebrew. 'Whole sheets of her ravings were written out, and found to consist of sentences intelligible each for itself, but with little or no connection with each other. Of the Hebrew, a small portion only could be traced to the Bible; the remainder seemed to be in the Rabbinical dialect. All trick or conspiracy was out of the question. Not only had the young woman ever been a harmless, simple creature, but she was evidently laboring under a nervous fever.' The case was followed up by a young physician, who succeeded in finding that at about nine years of age she had been living in the family of a learned Protestant pastor, who had had the habit of walking up and down a passage of the house which opened into the kitchen, reading in a loud voice out of his favorite books. A considerable number of these books were still in the possession of the old scholar's niece, who said 'he was a very learned man, and a great Hebraist.' Among the books were found a collection of Rabbinical writings, together with several Greek and Latin Fathers; and the physician succeeded in identifying so many passages with those taken down at the young woman's bedside, that no doubt could remain in any rational mind concerning the true origin of the impressions made on her nervous system. Coleridge adds, 'This authenticated case furnishes both proof and instance, that reliques of sensation may exist for an indefinite time in a latent state, in the very same order in which they were originally impressed.'

This hypothesis of a measurably permanent molecular arrangement of the cerebrum, as a basis of memory, seems, in some sort at least, to explain certain well-known phenomena of everyday experience. Every one has the power of overhauling his own knowledge in great degree at will; that is, of bringing back into consciousness what was once present in thought, but is now out of the field of view. This is commonly called the power of 'recollection' or 'reminiscence,' it being accomplished by a conscious effort of volition. If the physical signs of all such past acts of consciousness be still dormant in the brain, it only needs that they be revived to be re-read by the conscious self, and this may very well be effected by a reflex brain action, which sends (say) the ever varying blood-tides through the areas in which the molecular dispositions lie inactive, causing them to stand out, so to speak, like certain invisible inks, which, by the action of a liquid or heat, have their molecular arrangement so revivified as to render the dominant words distinctly visible. As to how the will acts to find the desired area, there seems to be no more difficulty than there is to explain how it is that it manages to find certain muscles of the body in speech and locomotion. One does not know in consciousness a single nerve or muscle of the many called into action to crook one's little finger, and yet it is no sooner thought upon than done. There seems to be no good reason why some analogous reflex mechanism may not exist to govern past knowledge. It is, perhaps, but a reverse action of the 'inhibitory' mechanism.

This purposive power to recall past concepts enables one to wander back into past scenes and 'fight one's battles o'er'; but its work is immediate and constant. It is necessary in every act and in every concept, for it is needed to recall the mental state of a moment ago, as much as years gone. It is absolutely necessary in every act of perception, for, since perception is not a single sensation, but a composite of many, this purposive memory must bring the divers elements together. This is of itself reason enough to establish the fact that memory has a physical basis, so long as such basis is granted at all in any psychic phenomenon.

Again, it is well known that at the most unexpected moment, and without any sort of conscious volition, the memory of something long out of the field of consciousness suddenly appears once more. Very often the recollection seems to stand alone; that is, out of any train of preceding ideas; but once present, it is clearly seen to connect itself with other memories as old as itself. Then, again, everybody knows that thoughts and memories are often forced upon one, especially when one wants to lose one's self in sleep, and cannot, or when there is a subject which one wants especially to keep out of mind. The theory of a physiological basis affords a sufficient explanation of these phenomena, since it is only necessary to assume that latent molecular arrangements are quickened into renewed activity by brain-stimulation of one sort or another.

Great stress has been laid upon what is called the "Law of Association," and the Empirical school of thought, from Hobbes to Spencer, attempts to explain by it the whole world of mental phenomena. It is by no means modern, but it has been enormously elaborated in these last years. The law may be stated somewhat as follows: Ideas or notions in the mind never stand isolated or single, and the presence of any one has a tendency to call up any other of the group to which it belongs. This seems an obvious and natural truth in view of the facts, as they have presented themselves to us, in the knowledge-forming and knowledge-conserving process. Association seems nothing more than the continuity of the empirical ego, with the brain-mechanism as its physical basis.

CHAPTER XIII.

THE IMAGINATION.

Definition. Classification. Cognitive and Sentient Imagination. Economic and Rational Imagination. Artistic and Rhythmic. Music. Relation of Memory and Imagination.

THE memory has no power to carry us back actually into the past, though it deals exclusively with acts of past consciousness, that is, with whatever now remains in the self of such previously excited activities. It projects these back into what we call time past, just as perception in time present projects its presentations into space.

Now the notion of past time carries with it, necessarily, as a negation, time to come or the future. The present is the meeting-place of the past and future; or perhaps better, they are but negations or limits on either hand of the present.

As memory is an activity or power of the self which deals with what *has* been, so there must be some activity which shall concern itself with what *may* be. The manifestation of this energy of the self is what we call the Imagination. It may be defined to be that power of the self through which past concepts, so modified and combined as to present some new and original element, are made to stand out with more or less vividness as objects of thought. If past scenes are returned to thought simply as they were (if that be possible), it is an act of memory; but as there is always some modification or change in the re-presentation of mental states, the imagination plays a large part in what is commonly set down as the unaided work of memory. Indeed, the functions of the imagination embrace the whole scope of mental action. Whatever is thought out as an end, in any sense, is new; and must find its place in the sphere of the possible — the 'becoming.' It cannot be memory which carries the self on towards and into that which is not yet actual, but only possible or beginning to be. This is the work of the Imagination. Its importance can hardly be exaggerated.

There is considerable difference of opinion as to the proper classification of the several domains of this power. It should seem that a scientific classification would be according to the following scheme : —

SCHEME OF IMAGINATION.

IMAGINATION
- Cognitive
  - Economic = Causality = Understanding = Utility.
  - (Ground) (Domain) (End)
  - Rational = Causality = Pure Reason = Truth.
- Sentient
  - Artistic = Space = Vision = { Painting. Sculpture. Architecture. }
  - (Ground) (Domain) (End)
  - Rhythmic = Time = Hearing = { Poetry. Music. }

By the grand division into 'Cognitive' and 'Sentient' Imagination, we cover the two classes which embrace all Knowledge, Intellect and Feeling.

The 'Cognitive' Imagination breaks up into two sub-divisions, one with causality as its ground, the understanding as its proper area, and utility as its end. It may be called, therefore, the 'Economic' Imagination. The other has causality for its ground, the 'pure reason' for its area, and truth for its end. It may be called the 'Rational' Imagination.

The 'Sentient' Imagination also breaks up into two subdivisions: first, the 'Artistic,' with space as its ground, vision for its domain, and painting, sculpture, and architecture as its ends; second, the 'Rhythmic' Imagination, with time as its ground, the sense of hearing for its domain, with poetry and music as its ends.

Under the 'Economic' Imagination we class all those activities of this power which have the useful in any form for their object. It embraces the widest possible range, starting from the level of 'every-day's most quiet need,' and rising to the highest possible pitch of inventive genius. The whole world of mechanical contrivances are its products. Any man who proceeds to do a new thing, — new not as unlike what has been done before by others, or even by himself, but new in that he is striving to cause that to be which otherwise would never come into being, is exercising a power which transcends the sphere of the actual and deals with the possible. Thus the savage who puts together two fagots to start a fire, or sharpens a stick for a plough, is an inventor, and in so far exercises the economic imagination. Every make-shift, shortcut, or contrivance has an element of originality in it, and is a protrusion into the future, making that actual which before was only possible. The construction of any sentence is an invention. Even if the thought be not new, there must be an arrangement of words and sentences which is new at least to the thinker, and the power which leads to this arrangement is neither that which discovers the actual to sense-perception, nor brings back the past in memory. It is a movement which peers into the future, and strives to evolve reality from the potential. The economic imagination has the useful in its various forms for its characteristic, — the useful in its broadest sense; so that all domestic, social, and political aspects of life in their practical applications, must depend upon this transcendent power of the personality as a necessary factor.

There is still another phase of mental effort, the broadest of all in its sweep, which, it should seem, must find its place under the economic imagination, and that is the concept-forming process. The philosophic world has rung with controversies as to what the content of a concept really is, but without entering the arena which has witnessed such valiant feats between Realist and Nominalist, we may venture to affirm that there is always some sort of shadowy and plastic image floating in the mind when a general term, such as 'man,' 'house,' or 'horse' is used, and that there is an element of originality in its structure. It is now pretty well agreed that there is no specific reality answering to such an abstract notion, and that it can have no proper image or simulacrum, but the power through which it gains whatever belongs to it, must fall under the head of the imagination in its sphere of use and necessity.

What we call the 'Rational Imagination,' is that mental capacity through which the evolution of truth as truth is evolved, and is founded — as we shall see all thought must be — in the fundamental pre-suppositions of what is called the Pure Reason. Here we have no longer the mere practical end which characterizes the economic imagination, but the motive and object is the evolution of truth, as truth. This is the domain of the mathematician, logician, physicist, theologian, and philosopher of every name. Every man is compelled to philosophize in some degree, ranging from cogency to fatuity, and so to every man must be conceded some degree of Rational Imagination.

There is one phase of this power much spoken of, but which differs in no essential particular from any other phase of the rational imagination, and that is what is called the 'Scientific Imagination.' That this power is necessary in any real work in science is too obvious to admit of question. D'Alembert said long ago that, to the geometer who invents, the imagination is not less essential than to the poet who creates. All hypotheses and theories are exemplifications of this constructive and original power. Illustrations could be drawn from every department of science, and from the humblest to the highest worker. Perhaps the most famous example is to be found in Kepler, who, by the power of a well-directed imagination, guessed out the primary laws of the solar system.

There might well be made many sub-divisions of the Rational Imagination, as well as of the Economic, but it is not necessary for the purposes in hand. It goes without saying, also, that numerous cross-divisions would be necessary to make this proposed scheme exhaustive. Indeed, it must be understood that the different phases of the imagination must overlap each other, and be variously combined; as, for example, in the drama, we have both the artistic and rhythmic phases distinctly marked. But we cannot dwell upon this.

The second great division of this original power which we call the 'Sentient Imagination,' covers the domain of what is commonly called Esthetics. It also breaks up into two principal divisions, one, objective, having a content which is susceptible of spacial expression, and so of appeal through the eye; and the other, subjective, with time as its ground, and the ear as its medium of recognition.

The 'Artistic Imagination' is not confined, of course, to the actual production of works of art, but includes all such exercise of the creative power as might be caught and delineated in color, in light and shade, or in plastic form by an artist of sufficient power. This pictorial power is possessed in vastly varying degrees, but there is no one who has it not in some measure. The power to recall the look of things implies it, since memory is rarely or never able to reproduce a sense-perception in perfectness of details, and these must be supplied by this artistic imagination. The whole world of art owes its existence to the distinctness with which the painter, the sculptor, and the architect is enabled to see in the mind the beauty of form, the charm of color, the grace and massiveness and power of his ideal, before it is caught in the meshes of the actual.

The world is not less indebted to the creative power of the Rhythmic Imagination. All elevated thought and speech, the power and beauty in the numbers of the poet — all arrangements of language in which originality and grace, sentiment and pathos, beauty and sublimity, are found single or combined, involve a rhythmic element, which, while not the sole ground, cannot be removed without fatal loss.

In music the rhythmic effect is still further heightened. The auxiliary elements which enter poetry through the concept-world drop out, and the ear is left free to revel in the possibilities of infinitely varied numbers. In melody the gradations of intervals, and successions of cadence, with endless play, rise and swell in adagio or andante sweetness, in the graceful flow of the legato, or the clear sharp ring of the staccato movement. The power and depth of harmony, the 'concord of sweet sounds,' opens a still further world of rhythmic power.

There is nothing more amazing than the grasp of the imagination in music. To hear in fancy hundreds of instruments and voices, blended and distinct at the same time, and follow them through the sweep and crash of, say, a Wagnerian opera, requires a power which is bewildering and mysterious.

Memory and Imagination are intimately related, but quite distinct in character. It is the business of memory to reproduce faithfully and servilely the actual; imagination rises above the trammels of dead reality and deals only with the possible. Memory speaks only as a witness; imagination is free, and creates for memory to report. Memory is the patient drudge; imagination is the master and lord; and, like a master, it uses the memory at every point. It is the original and creative energy of the self. It is only bound in that it is compelled to use the material furnished by memory. It has no power to create *ab initio*. The self cannot imagine a new sensation of any sort, — a new perfume, a new tint, or flavor, or any quality or power of the actual world. For all these, it must rely upon the memory. It has the power, however, to tear apart, and recombine and arrange. The result is new, and every effort which is a self-determined readjustment of existing concepts, is original. The Cyclops and the hippogriff were creations in the mind of some genius of antiquity in whose fancy the conceits first took shape. So with the centaur, the mermaid, the dragon, " gorgons, and hydras, and chimeras dire." These were all creations of the human mind ; and yet there is not a single sentient element in them all which was not supplied by an act of memory.

But though it must be admitted that this sphere-cleaving Pegasus is at the same time a quiet, domestic drudge, working by the side of memory, there is still this difference : imagination always has its face towards the pregnant future, while the memory looks back upon a dead past.

CHAPTER XIV.

DREAMING. — SOMNAMBULISM. — HYPNOTISM.

Phenomenon of Dreaming. Sleep. Do we always dream in sleep? The brain a thought-machine. Consciousness a mere phenomenon. The brain in sleep. Masso's observations. Character of dreams. Nightmare. Somnambulism. Case of student at Amsterdam. Case recorded by Dr. Abercrombie. German monk. Muscular feats. Double consciousness. Case of young lady at West Point. Hypnotism. Muscular effects. Dr. Charcot quoted. 'Suggestions.'

THERE is perhaps no physio-psychic problem which has proved more baffling in all attempts at explanation than the phenomenon of Dreaming. It has engaged the attention of philosophers from the earliest ages, and has been much discussed in the light of modern physiological research; but it still presents many difficulties.

The question which meets us at the threshold of the inquiry is one strictly physiological: What is sleep? It seems to be a universal phenomenon of animate nature; and is undoubtedly a periodical rest demanded in all vital action. Even the vegetable world shows states of repose; and in the lower animate world, as in insects, crustaceans, fishes, and reptiles the motionless state constantly recurring is hardly susceptible of other explanation. In birds and the lower mammals, the phenomenon admits of no question whatever. That it is a rest of the nervous system, with a renewal of the energy expended in the hours of wakefulness, seems clear enough; and the need of such refreshment is easily comprehended by man, who finds himself fatigued after periods of effort; but this is a pathological need read into the phenomenon, and affords no explanation whatever. It can hardly be expected that any perfectly satisfactory explanation of dream-phenomena will ever be reached until physiology can tell us what sleep really is, since dreaming is so closely related to this phenomenon.

The further question presents itself, and has been much discussed: Are we, when asleep, always dreaming? Opinion seems about equally divided, — at least the names on either side are sufficiently famous — the metaphysicians, as a rule, holding the affirmative, and the physicists the negative; — Descartes, Leibnitz, Kant, Hamilton, and the *a priorists* generally on one side, and Locke and his followers on the other. In the nature of the case, the question can never be fully settled. The only witness that can give first-hand testimony is consciousness itself, and it can answer only from memory: and as there are confessedly vast areas about which memory is a blank, it can never be finally known whether in those areas there is a continuance of consciousness or not. It seems to be certain, however, that there are often intervals of consciousness in those periods which at first seem blank, since all of us find ourselves remembering in the course of the day fragments of dreams of which in the morning we had no recollection whatever. It is also urged with strong probability, that as during our waking periods much goes on of which we are certainly conscious, and yet of which, a short time after, we have no recollection, the failure of memory to reproduce all dreams is not a sufficient reason for denying a continuity of consciousness. Again, it is urged that dreams are only the play of cerebral activity in intervals which are not properly sleep, but of partial wakefulness; and that they may be remembered or not according to the vividness of such action. The extreme rapidity with which dream-scenes chase each other — as, for example, in the case of Lord Holland, who fell asleep while listening to a reader, had a long dream, and yet awoke in time to catch the end of the sentence begun before sleep overtook him, — makes it possible to do any amount of dream-work in very inconsiderable periods of partial wakefulness. But the difficulty here is that many sleepers are dreaming — even talking and acting while profoundly unconscious of everything about them — sleeping on though roughly handled, and yet remember nothing of their visions, though they make the effort the moment after waking. The same thing is seen in the dumb animals. We have all seen dogs evidently engaged in the chase while apparently in the soundest slumber.

It is not to be doubted that even in the waking state there is a vast area of psychical phenomena which lies out of the circle of consciousness. This circle is extremely variable. When the attention is sharply fixed the field of consciousness is very narrow; but when there is a rapid change of attention from point to point it becomes indefinitely widened; so that there are at all times many reactions in the cerebral hemispheres which would be knowledge if taken account of by the self. The molecular changes which give rise to thought go on doubtless in sleep much in the way they do when one is awake, except that they are in large degree without purposive direction.

We are, indeed, driven to the conclusion that the cerebral hemispheres are simply thought machines. This does not mean, however, that the brain thinks: no machine ever does anything of itself. A machine is an instrument merely; and the idea of an instrument carries with it necessarily a higher power which uses it. For example — a loom is a machine by means of which certain materials are woven into a textile fabric: but the machine by no means makes the cloth. It does a certain work in putting the materials together; but there are two very important factors which it is wholly inadequate to furnish. They are, first, the materials wrought upon, and second the personal factor by which alone the machine and the materials are brought together. No mechanism, though carried to the utmost limit of perfection, could make cloth. If in the category of its perfections it be demanded that there shall be the power to go forth and gather its materials, discriminate as to quality, devise the pattern of the fabric and enjoy the prerogative of conscious self-regulation, the mechanism manifestly ceases to be a mere machine : — it becomes just what is meant by personality. Nobody ever yet saw anything done, that is, a purposive end accomplished, *when he was in the secret of the doing*, in which he did not know a personal element to be present. Iron and brass and wood do not rush together and form a steam-engine, but they are consciously combined by a thinker; and so of every possible contrivance. If it be answered that this is not true in the processes of nature, the obvious reply is that that is the very point in issue: that it is a pure assumption, in the face of all analogy, that there is no thought power by which its processes are directed.

In granting, therefore, that the neural mechanism in its highest form is a thought producer, the necessity for the material of thought (stimuli), and the presence of the operator (ego) must be presupposed. Now in many of the cunningly devised machines, the material having been selected and placed, it is possible for the operator to turn his back for a time, or even attend to something else quite different from the mechanical work, while it goes on perfectly well. This seems to be the case with all the reflex or automatic activity of the human mechanism in which at the beginning attention was necessary; but which, after being rightly set, goes on measurably well, even better sometimes without attention than with it.

Now it is hardly to be questioned that in the human thought mechanism changes and modifications are constantly taking place far below the threshold of cognition, and especially must this be the case in the silence of sleep. It is doubtless due to this fact that perplexing questions which have been undergoing a mental discussion within us, and have been left in an unsettled state before going to bed at night, appear to us in the morning, after an undisturbed rest, in a clear and decided light. Concepts have been brought together and arranged by the apperceiving power, which is an ever-active propulsion of our cognitive nature, and the result is a clearing up of what before was uncertainty and confusion. Cases of such unconscious cerebration must have fallen within the experience of everybody.

If what has already been said on the subject of the necessary co-action of the imagination in arriving at any conclusion be true, this cognitive power must have exerted an energy in such cases as these; and there can be no just ground for denying large possibilities to its action. Such action is what we call dreaming. If the repose is complete, there is no recollection of what may have gone on; but the activity of the neural centres may be so energetic as to force upon the sleeper a quasi, or even vivid recognition of its activity; and this the memory may reproduce with greater or less distinctness, either immediately upon waking, or after an interval of time.

This explanation implies, first, that consciousness is by no means the self, nor the distinguishing and essential characteristic of the personality — that is, consciousness in its fully developed, or cognitive form. But this must be true, else the personality would suffer absolute and fatal breaks in its continuity; indeed, would be, not an entity at all, but a mere state or phenomenon, — consciousness itself being the witness. For in a state of coma, in swoons, in dreamless (so far as known to one's self) sleep, in the antenatal state one is compelled to confess complete blanks, so far as consciousness can testify. It is for this reason that we hold consciousness to be a phenomenon of personality, and have called it the illuminated circle of cognition. The thought-mechanism may be in a state of perfect suspension, in which case there can be no consciousness whatever, or it may go on, and yet not come within the field of conscious recognition; and in this case there is no consciousness of such mechanical action.

The second point is this — attention is active and implies energy. In the state of relaxation and repose which we call sleep, the conative powers lack direction and control; that is to say, the will is, so to speak, off duty or asleep. This is a most important and significant point. It must be remembered, as we have seen already, that the functions of the will are not only to promote thought and action by concentrating attention and directing movements, but that an equally important part of its work is inhibitory. Indeed, it must be apparent, that but for this inhibitory power, recently discovered by physiologists, there would be inextricable confusion, even to chaos, in cerebral action. This power, in a way not yet fully understood, lays its hand, so to speak, on the vibratory chords in the encephalic organ to silence and prevent their action. The fact of the existence of such a factor in the neural system goes far to confirm the independence of the personality, and its supremacy over the mechanism which nature has furnished for its objective manifestations. If it were the brain-mechanism alone which gives thought and activity, then no management ought to be discoverable: that is to say, resuming our illustration of the machine, there could be found no room for the presence and control of the operator.

There are certain other facts, recently brought to light by physiologists, which have an important bearing on this general question. Passing over the work done by others, among the latest and most thorough observations are those of Mosso. He made observations on three persons who had lost portions of the cranial vault leaving the brain exposed, protected only by a soft pulsating cicatrix — a man and a woman each thirty-seven years old, and a child of twelve. He succeeded in taking simultaneous tracings of the pulse at the wrist, of the beat of the heart, of the movement of the chest, and of the exposed brain. He showed that during sleep there is a diminished amount of blood in the brain, and an increased amount in the extremities. He showed, further, that there are frequent adjustments in the distribution of the blood during sleep, and that this distribution is easily affected by external stimuli, — thus, a stimulus to the organs of sense caused a contraction of the vessels of the forearm, an increase of blood pressure, and a determination of blood to the brain; and on suddenly awakening the sleeper, there was a contraction of the vessels of the brain, a general rise of pressure, and an accelerated flow of blood through the hemispheres. During sleep, a loudly spoken word, a sound, a touch, the action of light, or any moderate sensory impression modified the rhythm of respiration, quickened the heart beats, and caused an increased flow of blood to the brain. He found that during very sound sleep these oscillations disappear: that the pulsatory movements are regular, and not affected by sensory impressions.

Three periodic movements of the brain during sleep are fairly certain: (1) pulsations corresponding to the beats of the heart; (2) oscillations of longer period, carrying smaller waves, thought in a general way, to correspond to the respiratory movements, and (3) undulations, still longer, thought by Mosso to indicate rhythmic contractions of the vessels of the membrane (*pia mater*) which covers the brain. It seems certain, therefore, that during sleep there is a comparatively bloodless condition of the brain. An examination of the retina by the opthalmascope during sleep, shows the same thing. Thus it can hardly be doubted that at least one of the physiological reasons for dreamless sleep is a depleted condition of the brain, with a diminished stimulation, and on the other hand, for the active dream state, the continuance of blood in the vascular system of the brain, or its return, from one cause or another after sleep has set in.

While dreams are sometimes coherent and sensible, they are, as a rule, full of incongruities, and even impossibilities, as presented to the judgment by memory. They present every possible variety between order and delirium. There are notable cases on record of vigorous thought-processes carried on during sleep. Condorcet tells us that the solution of a problem he had labored over for some time ineffectually was worked out for him in a dream; Condillac declares that subjects which occupied his thoughts upon retiring were continued successfully while asleep; and Coleridge has given us a fragment of a dream-poem — Kubla-Khan — which he remembered upon waking from a sleep in his chair. These are among the most celebrated; but most people have had experiences not unlike these, though perhaps more modest in degree. Such cases of method and coherency seem to follow close upon trains of ideas prosecuted vigorously in waking hours, and the fair presumption is that the power of directing the dream-imagery in some low degree at least is still active.

These cases, however, are altogether exceptional. It more often happens that one wakes with a vivid impression of some wonderful dream thought, poem, invention, or argument, which under the 'dry-light' of the understanding proves to be a 'baseless fabric.'

The 'stuff that dreams are made of' is usually very queer, sadly lacking in coherence and cogency. Assuming the directing and inhibitory power to be dormant, while the brain-cells are in a state of activity, presenting to consciousness the material of thought without system and arrangement, the most fantastic and grotesque results are accounted for. The living and the dead would commingle, as we know they do in dreams, and that without the least incongruity to a power which, in itself, has nothing to do but take note of what is presented. There can be no absurdity, no moral quality, no surprise, so long as the understanding is inactive; and it should seem to be for this reason that even time and space forms are so constantly set at naught in dreams.

But yet it often happens in dream fantasies that some question as to the reality of the scenes presented arises in consciousness; as, for example, when one has a dream within a dream. Some years ago, if I may give my own experience, I dreamed that in making some excavations for the foundation of a building, a little negro boy, about twelve years old, was dug up alive and in perfect health. It was such a remarkable event that it excited great attention; and a number of gentlemen (so the dream went) set to work to investigate the phenomenon. The boy was put through a series of questions; and, as he had been buried before the late war, then happily over, he was experimented upon, in his ignorance of the changed relations, to test his memory and witness his surprise at the new order of things. It all seemed so extraordinary, that I (the dreamer), suspected the whole thing to be a dream; and accordingly went out on the street, and looked about to see if everything was natural, — spoke to several acquaintances, asked them whether I was awake or asleep, and, having thoroughly assured myself that I was fully awake, went back to resume the interesting investigation.

In this case it should seem that it was not really the understanding which suggested the doubt, nor to which the test was submitted; but that it was simply a fancied or imaginary understanding, and so of a piece with all the rest of the scene.

In what is called 'nightmare' the dream-state is complicated by more or less futile attempts at muscular activity, with a sense of horrible inability to escape some fancied bondage or impending danger. When the movement is a part of the dream proper, and remains purely imaginary, the motor nerves seem not to be affected, though in the case of the vocal powers they often really act without any sense of oppression. Many people, especially children, talk in their sleep; and it is sometimes the case that sleepers will answer questions put to them, with a certain degree of freedom. Indeed, it is not impossible to direct the current of dreams to some extent by stimuli judiciously applied, the reason for which is obvious enough from what has been already said.

In the phenomena of somnambulism we have a more pronounced phase of psychical and muscular action combined. The cognitive powers remain clear in certain particulars, and the motor action seems unusually perfect, while sense-perception, except in certain special phases, seems quite suspended. Talking in sleep is an incipient form of somnambulism; but in its developed form the sleeper rises, dresses himself, enters upon the execution of some purpose, often of a delicate and complicated character; and finally returns to bed, and in the morning has no recollection of his doings, except, perhaps, as a mere dream. Cases are on record of persons who have performed all sorts of acts in sleep; as of mechanics who have got up, and gone on with their ordinary work; musicians who have shown higher powers than they possessed in their waking moments, — the voice sweeter and more powerful, the instrumentation more firm and delicate. One man was known to saddle his horse, and ride to his market-place, miles away.

The somnambule possesses sense-perceptions, or, at least, powers of external recognition, which are wholly unknown to us in our waking hours. For example, there is some power which seems to take the place of sight. Dr. Carpenter tells us of the case of a student at Amsterdam, to whom a difficult calculation had been submitted by his mathematical professor. He had toiled at it for three nights in succession; and on the last night had been compelled to retire without reaching a successful issue, his candle having burnt out, with no other in reach. Upon rising in the morning, he was distressed at the thought of again disappointing his professor, who expected him to accomplish the solution, when, on approaching his table, he found the problem fully solved in his own handwriting, and in what turned out to be a new and shorter way than before known; and all this he had done in the dark.

Dr. Abercrombie recites the case of an eminent person who, having been consulted in a difficult matter, studied it deeply for several days without arriving at definite conclusions. His wife saw him rise in the night, go to the writing-desk, in the same room, and write a long paper, carefully fold and put it away in the desk. In the morning he told his wife that he had had a remarkable dream, — had dreamed that he had delivered a clear and voluminous opinion in the case which had perplexed him for days, and that he would give anything to recover the train of thought. She directed him to the desk, where he found the opinion fully written out.

A case is on record of a young monk in Germany, whose work during the day was that of a scrivener. He often rose in his sleep and went on with his work. Divers experiments were tried upon him. Among other things a screen was interposed between his eyes and his manuscript while writing, and was found to make no sort of difference. He kept the right spaces perfectly, and dotted his 'i's' and crossed his 't's' with his usual accuracy. Indeed, the sense of sight seems not to be used by the somnambule. The eye is dull, and sometimes shut, and yet in walking, obstacles are avoided perfectly, although they are newly placed, as, for example, a chair put in front of the sleep-walker as he advances.

The muscular feats performed by somnambulists are equally unaccountable. Persons in this abnormal state seem to have no sense of danger, clambering out of windows, and walking on the narrowest ledges at giddy heights without the least sign of caution. They constantly do what they could not in a normal state. I may mention a case in point which has always puzzled me. When a lad I occupied the same room with a younger brother; he was about twelve years old. Upon entering the room one night I found him in his night-clothes, stretched at full length on the mantel shelf, with some books for a pillow. The shelf was quite narrow, and about up to his chin; how he got up I never could make out. There was no sign of chair or other thing to assist him. When I spoke to him he quietly landed on his feet and jumped into bed.

Another marvellous fact is that somnambulists sometimes have a sort of double consciousness, — in their waking state living one life, in their somnambulic state another, with an orderly sequence and coherence. This presents itself in many degrees and phases. Perhaps the most curious example is in the case preserved by Dr. Abercrombie in his "Intellectual Powers," of a young lady at West Point. The affection began with an attack of somnolency which was protracted much beyond the usual time. "When she came out of it, she was found to have lost every kind of acquired knowledge. She immediately began to apply herself to the first elements of education, and was making considerable progress, when, after several months, she was seized with a second attack of somnolency. She was now at once restored to all the knowledge which she possessed before the first attack, but without the least recollection of anything that had taken place during the interval. After another interval she had a third attack of somnolency, which left her in the same state as after the first. In this manner she suffered these alternate conditions for a period of four years, with the very remarkable circumstance that during the one state she retained all her original knowledge; but during the other, that only which she had acquired since the first attack. During the healthy period, for example, she was remarkable for the beauty of her penmanship, but during the paroxysm wrote a poor, awkward hand. Persons introduced to her during the paroxysm, she recognized only in a subsequent paroxysm, but not in the interval; and persons whom she had seen for the first time during the healthy interval, she did not recognize during the attack."

It is not to be disputed that there is much so far unexplained in the somnambulic phenomena. Indeed it is but one of a number of classes, of psychical as well as physical phenomena still remaining in the domain of mystery.

Another very remarkable state, first called, in modern times, Mesmerism, and later Animal Magnetism, Electro-Biology, etc., but now almost universally known as Hypnotism (from ὕπνος, sleep), is clearly allied to that of Somnambulism. Its present name was given it by Dr. Braid, of Manchester, who, in 1841, set out to show the Mesmeric craze, then extant in England, to be founded in fraud and delusion; but he soon found that it had an extraordinary psychic truth at bottom. He stripped it of the magical and magnetic elements, by showing that the artificial or pseudo-sleep could be induced in the simplest possible manner without passes, or magnets, or darkened chambers. His method — which still prevails — was to cause the subject to gaze fixedly upon some bright object held at from ten to fifteen inches from the eyes and a little elevated: in a short time the pupils begin to relax, the eyelids become unsteady, then close, or if they do not, the operator, marking the change which comes over the subject's expression, closes them with a gentle pressure of his finger, at the same time quietly stroking the brow or cheeks. A good subject, after having been hypnotized a number of times, may be put to sleep by gazing a moment at the point of the operator's finger, or by a look, or, as it is asserted, by a mere effort of will on the part of the operator. It is necessary as a rule that the operator shall have the good-will of the subject; but Dr. Charcot asserts, in a late number of *The Forum*, that he has often succeeded in surprising unwilling patients into the hypnotic state by suddenly disclosing an electric or magnesium light, and also by the use of a very large tuning-fork operated by an electro-magnet, gradually brought up to its full intensity, or by the sudden bang of a gong. He says, however, that these methods are not always successful; and it seems that the patients experimented upon were of a peculiar hysterical character.

Hypnotized patients lose almost all independence of will, and the imagination is in an analogous condition to that of the dream state. They obey any order, or suggestion of the operator, almost as if they were mere machines, and believe anything told them however absurd or ridiculous. Even the senses seem to undergo temporary change. They will eat or drink anything, however unpalatable, with apparent relish, if only told that it is something ordinarily enjoyable. A stick put into a patient's hands becomes a snake at the suggestion of the operator, and he drops it, and starts back with fright: a bundle of rags becomes an infant, and he fondles and caresses it: he will bestride a chair and go through all the motions of riding a horse, — nothing is too ridiculous; and even all sense of decency and honesty seem lost.

An extraordinary class of muscular effects is produced. A perfect rigidity or relaxation of a muscle or a set of muscles, or even of the whole body, can be induced by the operator. The arm or leg may be made absolutely inflexible, — the eyes fixed upon the ceiling in such a prolonged and constrained position as would be impossible in a normal state; the patient being left for an indefinite time gazing like a statue; or, the body in a state of rigidity may be 'laid like a log, head and heels on two chairs, so stiff and rigid as to bear the weight of the operator sitting upon him.' The patient may be kept for hours in the hypnotic state, and then roused by a series of passes, or by holding him a moment and blowing gently in the face.

There are still further marvels of this abnormal trance-like state. Says Dr. Charcot: "Take one example among a thousand. I present a woman patient in the hypnotic state a blank leaf of paper, and say to her: 'Here is my portrait; what do you think of it — is it a good likeness?' After a moment's hesitation she answers, 'Yes indeed, your photograph; will you give it to me?' To impress deeply in the mind of the subject that imaginary portrait, I point my finger toward one of the four sides of the square leaf of paper, and tell her that my profile looks in that direction; I describe my clothing. The image being now fixed in her mind, I take the leaf of paper, and mix it with a score of other leaves precisely like it. I then hand the whole pack to the patient, bidding her go over them, and let me know if she finds among them anything she has seen before. She begins to look at the leaves one after another, and as soon as her eyes fall upon the one first shown her (I have made a mark upon it which she could not discover), forthwith she exclaims: 'Look — your portrait.' What is more curious still, if I turn the leaf over, as soon as her eyes rest upon it, she turns it up, saying my photograph is on the obverse. I then convey to her the order that she shall continue to see the portrait on the blank paper, even after the hypnosis has passed. Then I waken her, and again hand to her the pack of papers, requesting her to look over them. She handles them just as before when she was hypnotized, and utters the same exclamation, 'Look — your portrait.' If now I tell her she may retire, she returns to her dormitory, and her first care will be to show to her companions the photograph I have given her. Of course her companions not having received the suggestion will see only a blank leaf of paper without any trace whatever of a portrait; and will laugh at our subject and treat her as a visionary. Furthermore, this suggestion, this hallucination, will, if I wish, continue several days; all I have to do is to express the wish to the patient before awakening her."

Here the marvellous point is the recognition of the particular leaf out of a number entirely like it, and that so perfectly as to be able to turn it always so as to keep the imaginary photograph on the same side, and right side up. But, as Dr. Charcot says, we need not on this account call in any preternatural agency; a sharpening of the powers of sense, however, must be admitted; there can be no doubt that every bit of paper differs widely from every other, however like it may appear to our senses in a normal condition, and it is only fair to assume that the sensibilities in their abnormal state are able to note these differences. By what sense this is accomplished it is impossible to say, but there are facts which entirely warrant us in the conviction that at least the sense of smell is wonderfully sharpened — almost transformed. Thus, Dr. Carpenter says, that he has known a youth in the hypnotic state, to find out the owner of a glove placed in his hand, in a company of sixty, by the sense of smell, — scenting at each of them until he came to the right person. In another case, the owner of a ring was unhesitatingly found out, from among a company of twelve; the ring having been withdrawn from the finger before the patient was introduced. He also says that he has seen cases in which the sense of temperature was extraordinarily exalted. The increased delicacy and power of the sense of sight in the somnambulic state has been already spoken of.

It should seem that in the case of a hypnotic patient the will of the operator is in a large degree substituted for the will of the patient, and that the whole mechanism of the personality is under his domination. It is a dream-state, in which the suggestions of the operator are substituted for the chance stimuli which affect the sleeper. The most remarkable fact is that suggestions may be made to affect the patient hours, or even days after being relieved from the hypnotic state.

CHAPTER XV.

THE UNDERSTANDING.

A technical phase of Cognition. Faculty of Relations. Thought proper. The lower animals. Pain. Logical element in man. The Syllogism. Dictum of Aristotle. Deductive and Inductive Methods. Reciprocal processes. Hypotheticals.

LIKE all other activities of the self the 'understanding' is easily distinguishable after it has reached a tolerable state of development; but as we trace our way back towards the first elements of knowledge, its functions necessarily grow less and less distinct, until it, with all other self-modes, sinks back into its source and ground, the personality.

The understanding, however, is discoverable at the very threshold of consciousness, and is always present in any act or mode of conscious self-energy. It is the thought power, — the instrument of all knowing; and can only itself be known in the 'empirical ego,' through its own function of distinguishing itself from the other two co-ordinate and necessary modes of the 'self,' sensibility and will. Sensibility is not sensibility, volition is not volition in the sphere of consciousness, until revealed to us through the understanding.

It must be borne in mind that thought or cognition has a wider scope than 'understanding' in its technical sense, — that sense-perception, memory, and imagination all fall under the head of the cognitive energy.

When, however, the understanding has become sufficiently differentiated for recognition, we discover it to be the Faculty of Relations. Through its office the self notes consciously the likeness or difference between two concepts brought together for comparison by the conative power. We must distinguish between it, and the energy which brings the concepts together for comparison. Its business is to comprehend, — to see and note the congruence or incongruence of the concepts as presented. It deals with relations simply as facts, and has no power to change jot or tittle. It sees, if so be, that one stick is longer than another, that one weight is heavier than another, that the centre is within the circle, that the radii are equal, and any and all other relations which lie within its scope. It is not omniscient nor infallible; that is, it does not discover to us all the relations which subsist between the concepts compared, nor is it always right with regard to those upon which attention is directed. The number and nature of those discovered, and the accuracy of the result depend upon the degree of excellence in the particular understanding of the person exercising it.

It may be well to remind the reader that all this lies in the ideal world. The sticks as objects in the external world are not, and can never be, compared. The comparison is between the two ideas in the mind. Even if the two sticks are laid alongside each other, this is no comparison. They might lie so forever, and there would never be any comparison until the perceptions of them in the self-world are judged of by the understanding. The truth of the concepts themselves, and the clearness of their relation determine the excellence of the result. In other words the amount of mental energy of which any particular person is capable, and the effort he puts forth at the moment determine the worth of any judgment. The difference between any concept — itself a result of the apperceiving power — and a judgment, which is the result of comparing one particular concept with another, is clearly one of degree only. Into the concept, there have gone many conclusions of the rudimentary understanding. They have lost all traces, perhaps, of the individual acts which, in a quasi or subconscious way, have entered into their composition, and are now concrete wholes; but these primary acts of comparison were unquestionably the work of the same power which we now in its developed state call the understanding. Apperception and understanding are therefore essentially the same thing. All knowledge is the result of this power of judging. The relations of the actual, whether with or without physical basis originally, are the discovery of the apperceiving power; and when so discovered they are committed to the conserver of facts, the memory. These residual judgments, explicit and distinct, are constantly merged into other and new wholes, which are nothing more than higher concepts, and are committed to the memory in a new sign or name, as, 'home,' 'business,' 'government,' etc. This process never ends.

There is this difference to be noted, however. So long as the actual mergence is not accomplished, of course the new concept is not formed. When I say A is B, the A and the B stand out in their individuality; and although the content in the proposition is their agreement in thought, there is a constant play of the understanding through the conative energy, between the two extremes; and so long as this is true, it is not a concept proper, but a judgment.

We can proceed one step further in the process of building up knowledge. When judgments are brought together and discovered to have a ground of agreement, the conclusion must itself be a judgment, with subject and predicate, so long as the judgments stand distinct before the mind: but when they are 'arrested,' and sink out of sight, a long step is taken toward a residual concept which shall absorb the terms of the conclusion as well.

Thinking, in its usual acceptation, is the process of judging, and does not emerge in consciousness until after a large accumulation of elementary knowledge. From what has just been said, as well as from the whole trend of what has gone before, it must clearly appear that thinking, in this sense, is a very different thing from the all pervading thought power which we have seen to be fundamental in character. Knowledge of a low order thus precedes thinking proper.

In the elementary stages of knowing there is no conscious effort in the concept forming process. The action is carried on through the propulsion of a life power — an instinct of the personality which blindly propels the self forward until it has gained the power of self-direction.

Thinking, then, in this sense has in it a larger element of the will power, a greater effort of attention, than is found in the lower stages of knowing. Perhaps in a right sense, elementary knowledge can hardly be called thoughts. It affects the mind rather in a sensuous than in an ideal way. There is no power yet of abstraction, and so there are not yet any ideas proper. Everything is concrete, and the personality is controlled as if pushed or pulled from without; and yet we must recognize the inchoate cognitive element all through it.

Thinking, then, really begins for us, when the conscious effort of comparing concepts for purposes of judging emerges in us. There must be two conspicuous factors, the differentiated power of abstraction and the conscious effort of judging.

In this light, we can now answer, in a manner, the question everybody is asking: 'Do not the lower animals think?' Manifestly not, in this only right sense of thinking. They seem to have neither of these two necessary factors, — the power of holding a concept in its differentiated aspect as an act of consciousness, nor of exerting the conscious effort of comparison. They undoubtedly know many things, and they perform many acts which have the look of discrimination; but their knowledge lingers in the sphere of automatic determinism, and is to them as the push and pull of a power without and beyond them; and their apparent discrimination is the result upon them of this propulsion. The dog knows his master, but he does not know him as master. He looks to him for food, and even protection; but it is a sensuous knowledge far down in the region of sense-perception, as a smell, as a pleasing or a disagreeable effect of heat, or light. He shows evidences of gratification at the master's presence and pines and mopes in his absence; but it is probably of no higher character than is involved in his drawing near the fire for comfort, or withdrawing from it when too hot. The sight of a stick, if it has been the instrument of punishment in the hands of the master, causes him to cringe; and a gun, if he is a sporting animal, fills him with gladness. But of pain as pain, in any construed consciousness, or joy as an emotion, he knows nothing.

The whole life of the animal is largely, if not wholly, automatic. The apparent judgments which we so often witness are not the results of conscious reflection, but of the prevalence of the strongest impulsion at the moment. The dog is far from reaching the stage of conscious personality. There is no true reasoning, and this seems to be the necessary conclusion upon physiological grounds. Those creatures which show the highest power of adapting means to ends are those, not of advanced brain development, but of the lowest order. There is no comparison between the community life, and the constructive power of the ant or the bee, and that of the horse or dog; yet the differentiation of the higher nerve-cells of the brain in these is far beyond what it is in those.

Pain in the lower animals is a very different thing from what it is in man. There being no conscious personality, and no reflection, there can be no self-pity, and no construing of the physical derangement, which is properly only pain, when translated into thought. They show all the effects of pain as we know it, but so does one often under the influence of anæsthetics. The extraction of a tooth under chloroform is frequently accompanied by howls of apparent agony and exclamations pitiful to hear; but the dentist goes quietly about his work with no emotion or sympathy, because he knows the patient does not feel it.

If the animal world had the power of forming judgments and the capacity for self-knowledge, there would be no reason why they should not proceed to form language and to show an advance in the rational scale. It is not from the lack of the physical organs, as witness the parrot and all that class of birds; but it may be seriously questioned whether any bird ever attached the slightest meaning to words, though as clearly articulated as in human speech. But even if the tones of the human voice could not be simulated by the dog, for example, language could as certainly be composed of barks and howls.

It perhaps will be urged in objection, that animals do communicate with each other; — and undoubtedly they do; and those, let it be remembered again, of lowest powers do it most thoroughly. But this is not language in the human sense, any more than the solicitations of appetite or the reflex action in the leg of a decapitated frog are the results of thought. It is all in the domain of sense, meeting with perfect response, but not through the construing power.

Again, it is to be doubted whether any animal has ever been known to do a voluntary act in which judgment proper can be discovered. In the ape-tribe we should expect to find it if anywhere, but they fail to exhibit it. The dog or the ape will freeze to death before either of them will replenish a fire. The fuel may be at hand, the animal may have seen it put on the fire a thousand times, may have actually put it on itself, if so taught, and yet when it is to be done for a purpose, the animal never rises to the necessary height. If putting the fuel on the fire were the immediate cause of the heat, they doubtless would do it. They will open gates of intricate construction, and even show by their motions that they want their masters to do acts for them which they are unable to perform; but they must be acts which carry, or seem to carry the end with it. A dog with a bit of meat on a shelf too high for him to reach will starve, with the full knowledge that the food is there, before he will push a chair into position to enable him to reach it. This is too large a demand upon his construing power, or, perhaps, appeals to a power which he has not. It may be well to say, however, that all this is a mere question of fact, and that there need be no hesitancy in admitting, if the facts should warrant it, that dumb creatures have this construing or concept-forming power in at least a rudimentary form.

This logical power is discoverable in the lowest orders of men, and reaches, in ascending stages of development, to the highest. The boor or the savage who never heard the word 'logic' is a logician, inasmuch as he forms judgments, and exercises the power of self-determination in the light of conscious reflection. The whole world went on, and achieved astonishing results in art, in literature, and in architecture, waged skilful wars, and solved high problems in civil polity, before Aristotle discovered, and laid bare to the conscious thought of man, the laws of this mighty instrumentality. The highest analytical feat ever performed by the wit of man, perhaps, was his discovery and systematic statement of the laws of logical thinking. Very little has ever been added to the science of logic since it came full fledged from the hand of the Stagirite. The power is one thing; the explicit consciousness is quite another.

Though we can but glance at the subject, some consideration of the principles underlying the science of logic is absolutely necessary to understand the further evolution of the knowledge-forming process.

An argument formally stated is called a 'syllogism.' It consists of three propositions, two of them called 'premises,' and the third the 'conclusion.' The subject of the conclusion, that is, the concept of which something is declared, and commonly standing first, is called the 'minor term'; the predicate, or that which is declared of the subject, is called the 'major term.' One of the premises must contain the major, and the other the minor term, and they are called, respectively, the major and minor premise. The third concept entering the argument is called the 'middle term' and must enter both the premises. The middle term is thus the common ground upon which both the other concepts in the argument must stand wholly or in part, and these concepts must thus agree, in so far as they overlap in this process.

Let us take a simple concrete example:—

    All animals are mortal [major premise].
    All men are animals [minor premise].
    Therefore, all men are mortal [conclusion].

In the major premise the whole class, 'animals,' is declared to fall under a larger notion, 'mortal' (major term), and in the minor premise 'men' (minor term) is declared to fall under that notion which is itself contained in the major term: it follows, therefore, necessarily, that the minor is contained in the major term.

This may be made readily apparent to the eye. Take three circles, thus:—

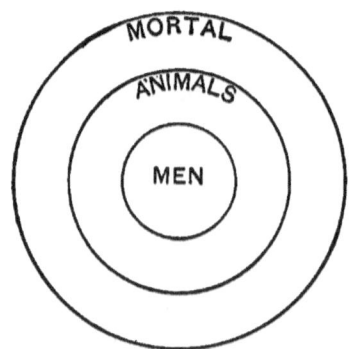

The largest circle is the major term (mortal), and contains the middle (animals), and this contains the minor (men); so that the minor term is necessarily contained in the major. This is a necessity of thought, and may be expressed generally thus : A part of a part is a part of the whole. This is in brief phrase the positive side of what is known as the Dictum of Aristotle. The Dictum itself — called the *Dictum de omni et nullo* — may be thus expressed : 'Whatever is predicated (*i.e.* affirmed or denied) universally of any class (*i.e.* of any whole) may be also predicated of any part of that class.'

To the test of this dictum, all deductive reasoning must at last submit ; and a part of the business of logic is to show how this test may be made to apply in all the possible forms an argument may assume.

In every argument there are two things to be noted, the 'form' and the 'matter.' By the form is meant all that is left after the content of the concepts is removed, the concrete meaning of the concepts being the matter. In our example, we have in the concept 'men,' a fasciculus of knowledge which has grown up in the empirical ego out of a multitude of experiences; and so in the other two concepts. Now, it may or may not be true that 'all animals are mortal'; or that 'all men are animals.' If we take out of these propositions all that is contingent (the matter) there will still be left the declarative element, and we shall have 'Something is Something,' or, 'Anything is Anything.' Let A be the symbol of anything, with the condition that it shall always remain the same whenever it may appear in the same argument; and B and C be symbols of like character. Substituting these in our example above, we have : —

All A is B.
All C is A.
Therefore, all C is B.

We have now eliminated all question of the truth or falsity of the declaration in the premises depending upon the matter of the concepts themselves, and have left the bare form of the proposition.

The conclusion now not only follows necessarily from the premises, but is indisputably true. In the concrete example, the conclusion, 'All men are mortal,' followed apodictically from the premises; but might nevertheless be false, because one or other of the premises might not, as a matter of fact, be true. Thus the formal syllogism in any of its various forms established as valid by the science of pure logic, gives indisputable conclusions; but as an empty form establishes nothing: the concepts must be restored in the argument before it is of any practical use.

We must thus have in every real argument the two factors, form and matter. The conclusion may be false, either from the fact that its particular form is not valid, or that one or both of the premises are materially false.

Pure, or Formal Logic looks only to the form, and has nothing to do with the matter. The truth of any premise is a question for Practical Logic; and at this door steps in the whole circle of the sciences. The question of the truth or falsity of the proposition, 'All men are mortal,' can only be answered by the science of biology. So in any proposition. Only Astronomy can tell us whether it is true that 'Jupiter is still in a semi-fluid state'; and we shall have to appeal to Chemistry to know the truth of the proposition, 'Water is composed of oxygen and hydrogen'; and so of all questions depending upon experience.

Now syllogistic reasoning is deductive in character; that is, it starts with a sumption (major premise), and the office of the *sub*-sumption (minor premise) is to declare that a certain subject falls under that sumption, or general rule; and the conclusion follows deductively. We thus descend from the general to the particular. The sciences arrive at their conclusions in a directly contrary way. They ascend from the particular to the general. They begin by noting a fact, and then another of like character, and so on until finally the number of accordant facts warrants the establishment of a rule, with greater or less probability as to its universality. This is called Inductive reasoning.

These two classes of the reasoning process are mutually dependent. There can be no induction without a *precedent* sumption; and there can be no deduction without a *precedent* sub-sumption. Every judgment is an inchoate syllogism. Take for example the declaration: Gold is yellow. The concept of the substance called 'gold' must be compared with the concept 'yellow,' and it must be seen to agree. The color seen in the gold is not the old or ready-formed concept 'yellow' in the mind; it is a particular or individual perception which the apperceptive power recognizes as agreeing with the already subsisting concept yellow. There are thus present all the elements of a syllogism; and it may be written out as follows: —

Whatever modifies light thus is yellow;
Gold has this property;
Ergo, gold is yellow.

No proposition can be made which will not fall under the same head. Facts are simply propositions, and therefore no fact can be stated which has not a syllogism lurking under it; and thus it is that the very elements of Inductive conclusions proceed from the principle of deduction.

But it will be asked: How then is the first fact of knowledge — the first element of the first concept had? I do not know that it would imperil one's reputation disastrously if one should frankly confess that one does not know. It would only go along with the answer to the question as to whence protoplasm comes, or whence anything is. The business of constructing the world out of nothing is not in especially high favor at this time; and we may respectfully decline to attempt it.

An answer may be given, however, in a way which without really explaining anything, carries us back as far as it is possible to go, and as far as we ought to want to go. It is simply this: Man is, by reason of his nature, a discoverer of likenesses and differences, and is thus an inchoate logician, and *must* begin to know from the first moment of his conscious existence. Thus we see again that logic as an act precedes by an incalculable span the science or systematic knowledge of logic.

That Deduction is in need of the help of Induction is freely admitted on all sides. Even in those cases where the premises are declarations of rational or intuitive truths, the Inductive element must be admitted, for no rational or necessary truth can be freed from an empirical element in the process of recognizing its axiomatic or universal character.

In these two much discussed methods of reasoning, we seem to have but the expansion of the necessarily reciprocal phases of all knowing, analysis and synthesis. These we have seen to be mutually necessary to each other. In them we found the time-worn problem of 'the one and the many'; and here we come upon it face to face once more. Of the principle of Induction we shall see something further as we go on.

We have spoken only of categorical syllogisms, — that is to say, of syllogisms in which the propositions fall under one or other of the four possible forms of unconditional or categorical propositions: All A is B, No A is B, Some A is B, and Some A is not B. In each of these forms there is simple affirmation or denial. But there are three other possible forms which propositions may assume. They are all conditional in character: the simple conditional, If A is, B is; the disjunctive, A is either B or C; and the dilemmatic, If A is, B is either C or D. The last, it will be seen, is a combination of the other two. It is not necessary for our purpose to enter upon any discussion of conditional syllogisms. The fundamental principle upon which they depend will sufficiently appear as we proceed.

CHAPTER XVI.

THE PURE REASON.

Intuitive knowledge. Conditions of explicit thought. Controversy about 'Innate Ideas.' Experimental knowledge. Law of Identity. Law of Contradiction. Excluded middle. Its questionable use in certain cases. Hamilton. Sufficient Reason. Causality. Hume. Locke. Leibnitz. The Laws of Motion. All science based upon necessary truths.

IN all processes of reasoning we deal with relations, or the bearing of one notion upon another. This is the domain of the understanding proper. It does not, in its practical operations, transcend the sphere of the limited, the conditioned, the finite. But relations presuppose a reality which underlies them, — a ground which enables them to be. The understanding takes things and events as they are presented to it; compares, arranges, and concludes, without necessarily asking why they are, or whence they come. We now seek the ground which supports the operations of the understanding.

It must be apparent that all along through the preceding inquiry touching cognition there has always been a somewhat taken for granted. Even at the very threshold of animate existence we found an element furnished to our hand, and without which science could take no step towards tracing the development of even the bodily mechanism. In the unit-mass of protoplasm, we found that we were compelled to recognize as pre-existing, the power of building up, or converting dead food into the living unit-mass — a power of self-movement, a capacity of responding to external stimuli through irritability or sensitiveness. In other words, we found life manifesting itself as a precedent condition.

This life with all that it involves, that is to say, all that is evolved through it, is a free gift. Although the effort has been persistently and repeatedly made, man has not been able to push this vital factor back into mere matter, or evolve it from dead matter. If it could be so evolved, the case would not be really different, since even then it would be only evolved; and the beginning, in the sense of the ultimate, would not have been reached, but only pushed back one further stage. Matter, instinct with life, could not be called dead, and even were it dead, the question would still obtrude itself: whence came this new factor which seizes upon and converts the material factor to its uses?

It is, then, manifestly impossible for a finite being to start without somewhat freely and absolutely given. But the original powers and capacities which the physicist is compelled to recognize as ready at hand in the protoplasmic mass are only the beginning of the gratuitously furnished capital which he is compelled to accept all along the way upward to man. He is compelled to recognize constantly emerging phenomena of which he is not at all in the secret. He cannot bridge the chasm between motion and sensation; he cannot point out even the direction in which thought lies from sensation, nor will from either. He finds himself in the light of consciousness, but he does not know when it came to him, nor how; he does not, and cannot, see it in another, and is as utterly in the dark as to what touch it has with the nerve-centres in the cortex of the brain, as our fathers were of its nature when they supposed it to reside in the pineal gland, or to exist independently of the body. It must be accepted as a revelation, happily bestowed at the auspicious juncture, no matter how far it may be made dependent, nor how perfectly synchronous with any stage of physiological development.

Now, since this is manifestly true, it is in no wise more wonderful that the conscious life should come ready furnished with laws of its being, than that the physical world should come to us with laws written in and through it. If it be answered that physical nature made its own laws, we ask: How do you know? They are unquestionably here, but nobody ever saw them come, and nobody knows what they are now that we see them. Who presumes to know what energy is? or chemical affinity? or motion?

This objection, then, is but an assertion without possible proof, and, as such, may be passed. But if it had all proof, the fact of the prior existence of these laws of the material world, as the reason and explanation of physical phenomena, will not be disputed by any.

In like way the actual existence of the laws of cognition cannot be disputed, nor their prior existence denied as the sufficient reason for the phenomena of thought. Whether these laws made themselves, is not the question. It is, Are they here? and are they logically prior to the development of knowledge?

Again, the question is not when they were first known to be dominating the knowledge-forming process. They might never be known, and yet be exercising their functions; just as gravitation did its work through countless ages, before man knew anything about it.

We have been stealthily approaching, in the preceding reflections, a venerable controversy; and we are now fairly on it. It is the question as to whether there are any '*innate ideas*' in the personality or not. In the first place, the name is an unfortunate one, prejudicing the question in the negative at the start. It may be safely asserted that nobody, fairly entitled to an opinion, holds, and that nobody, following Leibnitz in modern times, ever held, that the personality begins its work of gathering knowledge with a set of ready-made ideas in the mind; somewhat as a clothier might be supposed to commence business, with a lot of made-up garments in stock! That there are any innate ideas in this sense, no one fairly informed in the matter would presume to assert. To return, then, to the line we were pursuing: in the course of the development of knowledge, after the manner we have briefly sketched, there are discovered certain laws which govern psychical action. These are facts of the self: and they are not derived from the external world. Let it be carefully borne in mind that there is no intention in using the word 'derived' to deny that experience of the external world is the *occasion* of their becoming known to the self. If one should find one's self in a perfectly dark room full of divers mechanisms, one would see nothing of them; but if a light should be struck, they would be immediately revealed. The light would be the occasion or medium of their discovery, but not the cause. So of these original truths which belong to the self. Experience is the torch through which they are revealed to cognition, as well as the material upon which they operate. They would remain potentially in the mind forever, but could never become known to the self if no stimuli from the external world were ever presented to rouse our psychological energies. There never could be any empirical ego; and so the self would never know that it had in it the potentialities of sensibility and cognition, and the power of self-determination.

But, on the other hand, no matter what might be the number and character of the external stimuli, and no matter how continuously they might be applied, there would be no sort of experience if the psychological factor of the self were not present. It would be like one ringing at the door-bell of an empty house. There would be nobody at home, and no response. Thus, experience itself is absolutely in need of a conscious presence to lift its physical basis from the dead world of mere motion and mass, into the life-world, before it can be experienced at all.

Let us look at this a moment further. Take the illustration above, of the door-bell and the empty house. Suppose the ringing from without should continue, and that any number of the simplest, or the most intricate signals were made with voice, or in any other way: what would be the result within? The question hardly needs answer. If there were on the inside, plastic substances and sensitized plates to record all pressures and all effects of light, what would there be at last within, but so many pressures unfelt, and so many pictures unseen? There they might remain forever meaningless and dead. But now introduce an occupant. Those sounds and signals would become at once what they were not before. They would have meaning read into them by personality; they would become the physical basis of thought and feeling.

This illustration, if not made 'to go on all fours,' is a fair representation of the relative action of the physical and psychological factors in the simplest possible act of experience or sense-perception. The caution is especially intended to guard against the notion of the personality as shut up, or in any wise enclosed in the brain, or in any part of the body. As we have seen already there is no possibility of giving it a local habitat.

Now it seems clear that the self has the inherent power of giving to molecular motions one thing which they had not themselves, and that is *meaning*. It has the power of composing or uniting separate and several molecular acts into ideal wholes. But this is not done pell-mell and indiscriminately; but with method and purpose. It must, therefore, be granted the power of discovering likeness and difference; and this simple fact conceded, the gulf is passed between mere mechanism and personality; — over the gulf which has no bridge in thought, thought passes and repasses continually in fact.

It requires a little patient thought for us, accustomed all our lives to the matter-of-course use of this power, to see at first that there is anything to wonder at in such an elementary energy. But the more one thinks about it, the more impossible it becomes to say why A is A. One is disposed to exclaim, with some feeling of contempt: 'Because it is.' The answer is philosophically correct; but where does the 'because' come in? This word implies that we discover a reason; but if so, what is it? In our scorn we may say again: 'It has to be.' Yes; but why? Any number of varying forms of asseveration may be tried in turn, but no one of them will advance us one step further towards an explanation. There is no reason, except this simplest of all, *It is*. The difficulty in the case arises from its very simplicity and obviousness. We are compelled to see that a thing is itself, and that is the beginning and end of all efforts at explanation or comprehension of the fact. This is the first fundamental principle of all knowing. It is called the *Law of Identity*. It is certainly not derived from anything, because it is not separable into any possible elements; and it does not come to us through (*i.e.* is not acquired by) experience, since it is a condition precedent, in the first and simplest act of experience.

The principle of difference, or unlikeness — called the *Law of Contradiction*, is equally obvious and necessary. To say that A is not non-A (assuming of course that one first understands the use of such an abstract formula) needs no reflection and no instruction. It is, however, as the negation of the law of identity, dependent upon it. In strictness, identity and difference are not two distinct and several principles, but are simply the two indissoluble phases of the one act of limitation. Likeness is inconceivable without its correlation, unlikeness; and the converse. But identity can be rightly called the positive, and difference the negative phase, since a thing cannot be 'othered' — as the Hegelians say — until one, in some sort at least, identifies it as somewhat to be othered. The position of Fichte is indisputable, that the ego could never know itself as ego, if it were not that it finds itself limited by the non-ego. The negative character of difference seems equally clear, since no aggregation of mere differences would give 'thing' a content.

Again, of two contradictories, — that is to say, two concepts or judgments, such that they, taken together, comprise all being, the affirmation of either absolutely denies the other; and the denial of either necessarily affirms the other: thus A and non-A are contradictories, since non-A comprises whatever there is in the universe not found in A; and the converse. It is obvious that there is no middle ground between them; and this absolute exclusion is the third fundamental truth of logic, called the *Law of Excluded Middle*.

Now, this doctrine of 'Excluded Middle,' while undoubtedly valid in itself, is easily susceptible of abuse in practice; and has been often used to false ends. It is necessary that the antithesis shall be certainly established before an argument is based upon it. Even as great a logician as Sir William Hamilton has failed, it should seem, in his use of it. In his " Philosophy of the Conditioned," he uses the doctrine of Excluded Middle with respect to Space and Time. Thus: Space is either limited or unlimited. It is not limited, therefore it is unlimited. But there is grave doubt as to whether it is either one or the other, — many of the ablest philosophers holding that space is not an objective entity at all, but is purely subjective. But to make the point more obvious, let us take a manifest absurdity, thus: Mercy is either animate or inanimate. Suppose one could prove that it is not alive: are we therefore shut up to the conclusion that it is dead? If I may so say, the doctrine of Excluded Middle is a mill, but one must be sure of what is put into the hopper before one can be certain of what is ground out. Great circumspection is necessary whenever one attempts to put the Infinite, the Absolute, the Naught, the Ultimate, in any form through any sort of logic-mill; and it is not surprising that queer grists have been ground out, from time to time, by those who have flattered themselves that they had emptied one or more measures of these transcendental concepts into the hopper. It is not a mere figure of speech to call logic a mill; for it is, in itself, as much a piece of dead mechanism as a corn-mill or a sewing-machine.

Again, whenever a change takes place which we recognize by the law of difference or contradiction — whenever a thing begins to be or become, or an event happens, we, by a necessary law of the self, think there must be some reason or cause which produces such change or event. This principle of causal dependence is the fourth fundamental law of logic, called the '*Law of Sufficient Reason.*'

The question as to how man gets hold of the notion of cause has been long a subject of earnest controversy. Hume occupies a conspicuous position in the discussion — indeed, the battle seems to have been waged about him as the *point d'appui*. It was he who, as an antagonist, set the Scotch school upon its mettle; and the great philosopher of Königsberg tells us that it was Hume who first woke him to the fray. It is not my purpose to follow the history of the contention, or to attempt any detailed treatment of the subject; but the matter is too important to pass in silence. David Hume and his followers, James and John Stuart Mill, Herbert Spencer, and the whole school of empiricists, declare that there are no original or necessary notions in the mind; but that all knowledge is derived from without through experience. The reader will readily perceive from what has gone before, that in one sense the present writer has no objections to this statement. "*Nihil est in intellectu, quod non fuerit in sensu*" is certainly true, if it be understood in the light of the added phrase of Leibnitz "*praeter intellectum ipsum.*" Manifestly there could never be any material of thought in the mind if the senses did not do their office; but, as we have seen, if there were no mind to give it meaning, there would still be no thought.

This carries us back to the question as to whether the psychical nature of man has laws which pervade it, or, in other words, whether it has a nature at all.

Now Locke, who is commonly regarded as the father of modern empiricism (though both Gassendi and Hobbes must have precedence in time), is perfectly clear, though not always consistent in his recognition of the native powers of the mind. Almost at the beginning of his famous essay, in speaking of maxims, he declares that they are not founded upon reasoning, "for all reasoning is search, and casting about, and requires pains and application; and how can it, with any tolerable sense, be supposed that what was imprinted by nature, as the foundation and guide of our reason, should need the use of reason to discover it?" Although there is a confusion here according to modern terminology in using 'reason' and 'reasoning' as interchangeable, his meaning is that the process of reasoning must find a foundation in that which 'was imprinted by nature,' and which is now called the Pure Reason. In the fourth book of the essay there is constant recognition of this truth. He says: "If we will reflect on our own ways of thinking, we shall find that sometimes the mind perceives the agreement or disagreement of two ideas immediately by themselves, without the intervention of any other: and this, I think, may be called intuitive knowledge. For in this the mind is at no pains of proving or examining, but perceives the truth, as the eye doth light, only by being directed towards it. Thus the mind perceives that white is not black, that a circle is not a triangle, that three are more than two, and equal to one and two. Such kind of truths the mind perceives at the first sight of the ideas together, by bare intuition, without the intervention of any other idea; and this kind of knowledge is the clearest and most certain that human frailty is capable of. This part of knowledge is irresistible, and like bright sunshine, forces itself immediately to be perceived, as soon as ever the mind turns its view that way; and leaves no room for hesitation, doubt, or examination, but the mind is presently filled with the clear light of it. It is on this intuition that depends all the certainty and evidence of all our knowledge; which certainty every one finds to be so great, that he cannot imagine, and therefore not require, a greater: for a man cannot conceive himself capable of greater certainty, than to know that any idea in his mind is such as he perceives it to be; and that two ideas, wherein he perceives a difference, are different, and not precisely the same. He that demands a greater certainty than this, demands he knows not what, and shows only that he has a mind to be a skeptic, without being able to be so. Certainty depends so wholly on this intuition, that in the next degree of knowledge, which I call demonstration, this intuition is necessary in all the connections of the intermediate ideas, without which we cannot attain knowledge and certainty."

Many other passages could be quoted to the same end, but it is unnecessary. The confusion which seems to lie at the bottom of the endless discussion on the subject of innate ideas is the failure to distinguish between the native potency of the personality to discover necessary truths as occasion requires, and an explicit notion of such truths already subsisting in the mind previous to any possible experience. In this later sense, as Locke and Hume and all that school contend, there are no innate ideas; and this is only to say what a far greater than any one of them declared at the beginning of the controversy, — Gottfried Wilhelm Leibnitz.

Hume recognizes the inherent power of the mind to discover necessary truths as clearly as his great predecessor, Locke. "The mind of man is so formed by nature," that it sees and feels so and so; the belief that heat and cold will follow contact with flame and snow "is the necessary result of placing the mind in such circumstances. It is an operation of the soul, when we are so situated, as unavoidable as to feel the passion of love, when we receive benefits; or hatred, when we meet with injuries. All these operations are a species of natural instincts, which no reasoning or process of the thought and understanding is able either to produce or to prevent." Again, "Reason is nothing but a wonderful and unintelligible instinct in our souls, which carries us along a certain train of ideas, and endows them with particular qualities, according to their particular situations and relations." Speaking of mathematical truths he says: "Propositions of this kind are discoverable by the mere operations of thought, without dependence on what is anywhere existent in the universe. Though there never were a circle or a triangle in nature, the truths demonstrated by Euclid would forever retain their certainty and evidence."

Hume addresses himself especially to the question of causation, contending that all that experience reveals in two perceptions which appear to have a causal nexus is their succession, — that "the mind never perceives any real connection among distinct existences," *i.e.* never *sees* the nexus between them; and I fancy nobody who has really reflected upon the matter thinks he can. The point of the difficulty seems to be just here, and is precisely the same we saw a few pages back in the illustration of the door-bell. There would be nothing but succession (and even that would not be perceived, if the perceiver be not granted), and these successive existences could never have any thought-nexus except for the unifying power of the thinker. It is just this power which discovers itself to us in attention, and in every conative act. When we have got far enough to formulate the thought, 'every event must have a cause,' this ideally-real nexus has risen up in us as an object of knowledge, and we cannot banish it. We then see that if a thing be let alone, it will continue to be just where and just what it is.

Now Hume saw this distinctly, and expressed it unequivocally. "Upon the whole," he says, "necessity is something that exists in the mind, not in objects; nor is it possible for us ever to form the most distant idea of it, considered as a quality in bodies." Again: "As the necessity, which makes two times two equal to four, or three angles of a triangle equal to two right ones, lies only in the understanding, by which we consider and compare these ideas, in like manner, the necessity of power, which unites causes and effects, lies in the determination of the mind to pass from the one to the other." This is found in his "Treatise of Human Nature." In his more mature work, the "Inquiry concerning Human Understanding," which he expressly desires to supersede 'that juvenile work,' he makes the following emphatic statement: "It is universally allowed, that matter, in all its operations, is actuated by a necessary force, and that every natural effect is so precisely determined by the energy of its cause, that no other effect, in such particular circumstances, could possibly have resulted from it. The degree and direction of every motive is, by the laws of nature, prescribed with such exactness, that a living creature may as soon arise from the shock of two bodies, as motion, in any other degree or direction, than what is actually produced by it."

He does not consider that there is any conflict between these statements, but it is interesting to note that he avows that there is a causal necessity in the understanding, and an equally certain causal nexus in the material world. His point is that while this is true, the fact of this causal certainty in the objective world is established as a fact in the understanding and is due to experience; with which we have no quarrel, so long as it is borne in mind that experience is but the occasion of the discovery in consciousness of this necessary law of personality. That this characteristic of mental necessity could not be derived from an accumulation of mere experiences needs no better witness than Hume himself. He says: "Matters of fact, which are the second objects of human reason, are not ascertained in the same manner [as mathematical truths]; nor is our evidence of their truth, however great, of a like nature with the foregoing. The contrary of every matter of fact is still possible, because it can never imply a contradiction, and is conceived by the mind with the same facility and distinctness, as if ever so conformable to reality. 'That the sun will not rise to-morrow,' is no less intelligible a proposition, and implies no more contradiction, than the affirmative, 'that it will rise.' We should in vain, therefore, attempt to demonstrate its falsehood. Were it demonstrably false, it would imply a contradiction, and could never be distinctly conceived by the mind."

Now this notion of cause has just this characteristic, that its contradictory is absolutely unthinkable, and upon it the whole magnificent fabric of science rests. Mechanical action is, as we have abundantly seen, the final test of all scientific research. But the whole doctrine of mechanics rests upon the first of the three Laws of Newton, the second and third hanging upon the first. That law is a pure metaphysical statement, without the slightest possible experience to support it, and with all experience against it. It is, 'Every body continues in a state of rest, or of uniform motion in a straight line, except in so far as it is compelled by force to change that state.' But did any one ever see a body moving in a straight line with uniform velocity? There is no such thing as a straight line, and probably by no possible contrivance could a body be made to move accurately in a straight line. A body falling under the action of gravitation does not describe a straight line, but one of the conic curves, and the equation of a projectile only becomes that of a right line (a particular case of a parabola) under suppositions which never can be true.

Is not science with unbroken voice assuring us that there is no such thing as actual rest in the universe, and that there is no such thing as unimpeded motion? Who, then, ever had experience of either the one or the other? And yet, who that understands the proposition fails to see that its contradictory is unthinkable? It is the law of persistence. A thing cannot take on a new state unless there be somewhat to compel it; that is, there must be a cause, or there will be no change.

The second law declares that 'change of motion takes place in the straight line in which the force acts,' but no one ever *saw* a body deflected from its path move in the line of the deflecting force, but in the diagonal of the two velocities. While the law is necessarily true to reason, it is not true to the senses.

The third law is, perhaps, even more at variance with common experience. If a cart pulls back upon a horse just as much as the horse pulls forward, 'How in the name of common sense,' asks the untaught mind, 'does the cart get on?' If a boy and a man are tugging at a rope, the man pulls the boy towards him, and that certainly does not look as if the boy's end of the rope was pulled back with exactly the same force as the man's is pulled in the opposite direction; and yet it undoubtedly is; for, says the third law, 'Action and reaction are equal and contrary.' And thus it seems plain that we should never have had any science of mechanics if the world had had to wait for the senses to give us the fundamental truths upon which it is so firmly established.

Let it be granted, then, that the first light thrown upon the fact that cause is a necessary law of mental action, comes from the discovery of the precedent fact that we find our bodily movements to be the results of effort; and let it be further granted that we cannot *see* any nexus between the will, and the movement of the muscles (as is certainly true), and yet the fact remains that the whole world including Hume, thinks, and cannot help thinking, that everything which begins to be what it was not, must have something acting upon it, quite otherwise than as a mere precedent existence; that is, the whole world knows that every event must have a cause.

That energy of the personality which discovers to us *necessary* truths is called the 'Pure Reason.'

CHAPTER XVII.

EMPIRICAL AND RATIONAL TRUTH.

Conditional Syllogism and Law of Sufficient Reason. No law of the natural world above doubt. Not so in thought. '*A priori*,' 'original,' etc., truths. Necessity the characteristic. Relation of Deduction and Induction. The basis of Induction. Intuition of Space. The Infinite and Absolute. The 'Philosophy of the Conditioned.'

THE fourth fundamental principle of logic, the law of 'Sufficient Reason,' is founded in the notion of cause, and is that which gives validity to conditional syllogisms. If A is, B is, expresses a causal nexus between A and B, such that the existence of A, the antecedent, necessitates the existence of B, the consequent; and so the principle is sometimes called the law of 'Reason and Consequent.' The position of Hume with regard to the sequence of events, as we have seen, is perfectly true in the realm of things. A thing or event does not necessarily imply the existence of that which precedes it; *i.e.* change is not the effect of mere antecedence; and we can never know without question, that any effect in nature is compelled by that which appears to be its cause, and just for the reason Hume urges; namely, we cannot see the physical links in any succession of existences: but with concepts, connected by a law of the Reason, the difference is marked. In a hypothetical proposition, such as that given above, where *ex hypothesi*, A is the reason for B, the causal nexus is indisputably established in thought. If it be accepted that, ' If A is, B is,' a dependence is established between antecedent and consequent which is inevitable. This relation has, therefore, a certainty which never can be felt as between any two existences in the world of things.

If now it becomes categorically certain that 'A is,' *i.e.* if the condition be removed by the affirmation of the antecedent, the existence of B follows with a certainty greater than that the earth is.

The negation of the condition — that is, the denial that 'A is,' — carries nothing with it whatever; because there is no hypothetical relation implied between the non-existence of A and B; B may exist from a thousand other reasons than the existence of A. Take a concrete example: 'If the price of flour goes up, the poor will suffer.' Here it is obvious that though flour remain stationary, the poor may suffer from cold, or disease, or what not.

Neither is there any relation established affirmatively in the proposition between B's existence and A's; for B, as we have just seen, may exist from other causes; the poor may suffer though flour continue steady. There remains but one other way of treating the proposition. We can deny the consequent. In that case we touch the causal nexus, and the denial of A also follows necessarily. If the poor do not suffer, the price of flour cannot have advanced.

It is just because the causal connection between separate acts of existence can never be discovered, as between things and events in nature, that no law of the natural world, as Hume declares, can be raised above the possibility of doubt. We do not doubt, it is true, that ponderable bodies when released from their supports will fall to the ground, under the action of gravitation; but it is not unthinkable that they might be deprived of their weight suddenly by the action of some new Aladdin's Lamp, and fly upwards. The difference, thus, between necessary truth, in which the causal connection is in the form itself and not at all dependent upon the content, and empirical truth, in which the connection, being dependent upon the content, can never be seen, is world-wide. In the one class we cannot even start to doubt; and in the other the roads to doubt lead off in every direction. In all mathematical truths, doubt is impossible; to which Hume bears such unequivocal testimony. Try as we may, we cannot think that a part is as great as the whole, and so with all axiomatic truth, and all conclusions which flow apodictically.

Now let it be borne in mind that the proposition is not that all persons alike see truths which have in them this characteristic of necessity. It is often urged as an objection, for example, that in a column of figures several persons may find the sum quite different. Undeniably, one may make blunders in adding; but does anybody doubt that the sum is certain for all that, and that the trouble is in the unconscious slips in reckoning it? So with all necessary truths which are recondite and locked up to all but those of high mathematical or logical perception. Truth in its very nature is absolutely definite; but it is not always axiomatic. When not at once apparent we approach it from the axiomatic side, and the results emerge deductively and necessarily, unless — and this, as we shall see further, is an important exception to note — unless we at some stage of the deductive process introduce concealed or pregnant factors; in which case, the conclusion may be worthless. The apodictic or inevitable character of every stage of an argument, and of every element introduced, is presupposed.

Axioms or self-evident truths, then, are the original elements out of which, or in consequence of which, all conclusions of a deductive and necessary character result. Even when they are simple — that is, cannot be resolved into any possible elements — they are not equally conspicuous to all persons, nor to the same person at all stages of mental development. They are not seen at all by any in the infantile period, nor by those of low mental development at once when pointed out; and yet, in point of fact, they are actually employed by every one long before they become articulately differentiated in consciousness. Their self-evident character consists in this, that when once brought into the light of consciousness as objects of thought, they cannot be resolved into more simple elements, and have no ground deeper than pure necessity; their contradictions are, therefore, unthinkable. Many people live and die without knowing what an axiom is, or so much as that there are any; but they, all the same, have been acting upon them practically at every stage of their conscious life; and can be made to recognize the fact if their attention be properly directed.

Take an illustration of this, though the point is almost too plain to require it. Suppose the veriest boor, without any sort of education, and without much native mental vigor, be asked to find out whether two posts in the ground are of the same length. He would most likely take a rod, and mark the length of one of them on it, and then try the other by it. Nobody could do better; and if his work is done carefully, the result is inevitably correct. Now why is it that these two posts are pronounced equal in length (if it should so prove) through the medium of the rod? The self-evident truth upon which the whole action is based, when brought to the light and clothed in words is, 'Two things which are equal to a third are equal to each other.' It is the same truth which we saw underlying all syllogistic reasoning and all judgments whatever. It is necessary for the first act of comparison, and is therefore presupposed from the very beginning of all accumulated knowledge. It is a law of the psychical factor of the self; and such truths are called, '*a priori*,' 'original,' 'primordial,' 'intuitive,' 'fundamental,' 'necessary,' and, I do not know by how many other names,—all meaning substantially the same thing. They belong to that phase of our cognitive nature called the 'Pure Reason.'

A distinction is, then, to be made between the Understanding and the Reason — a logical distinction, though by no means a separation. Through the understanding we put things together, or separate them into classes or units, — we synthesize and analyze, — but these operations have their ground in the deeper truths of the Reason. Reason and Reasoning are therefore to be distinguished in thought. The 'Pure Reason' is not properly the domain of thought at all, — it is the foundation upon which the whole thought fabric rests. It is itself blind, and so does not see; but the mental seeing or thought power is dependent upon it for its validity in every act. It is founded in the psychical sense-world, if I may so speak, and is to the higher modes of the self in the discovery of apodictic truth, what sensation, with its physical basis and dependence upon external stimuli, is to the world of sense-perceptions. Not that it does not reach down into these as well; because it is, so to speak, the final court of appeal in all possible knowing, and its touch must be co-extensive with the whole personality. There is thus the very best ground for calling the presentations of the pure reason intuitions, or face-to-face knowledge. They, as making all hyper-physical knowledge possible, should seem to be analogous to the senses which make all bodily knowledge possible. They have their roots in sense-perception, much as the senses have theirs buried out of sight in sub-conscious sensation; and there thus seems to be two measurably distinct psychical worlds, — one the sense-world, based upon the immediate presentations of the physical organs, with a very slight intellectual factor; and the other, the rational world, distinctively ideal, based upon the face-to-face presentations in the Pure Reason. The one is the region of brute-life; the other of soul-life. In man the two are conjoined, — the ideal mode superposed upon the animal mode with the power to sublimate and refine the lower to its higher uses.

We have seen that the distinguishing characteristic of rational truth is necessity. In the conclusion based upon sense-perception, or contingent truth, the highest reach is probability, and the chasm between the highest probability and a necessary truth cannot be spanned.

Besides this, or, better perhaps, in consequence of this, the method by which any contingent truth is reached is the reverse of that leading to a necessary conclusion. In the case of necessary truth the results or conclusions flow out from the basic elements given in the intuitions or axioms of the pure reason, as water from a fountain. In the case of contingent truth the results are an accumulation, and are as a reservoir filled by the flow from the living source. This last process is Induction. It is, as indicated above, synthetic; the rational process being analytic.

Now, if I may use my figure a little further, in the reservoir there is always something more to come, and this goes on forever; while in the flowing stream, it is just what it is, from any given source. Synthetic or contingent truth is as the water in a reservoir of infinite capacity, always increasing, but never full; analytic or rational truth is living reality, exact and definite, though different persons see more or less of it from any particular point of view.

The inductive method is to add fact to fact until a probable conclusion is formed; and as the number of facts increases, the conclusion becomes more and more probable, becoming at last overpowering; but never acquiring the quality of necessity. For example, some one in the beginning observed the fact that all the horned animals he knew were ruminant. Others observed the same thing; the inquiry was still further extended, until all parts of the world had been explored, and the facts always remaining the same, it has become an established truth which no body doubts. But one can think the contrary easily enough, — think that it might not be so; and, more than this, it may not in fact be true after all; for there are still parts of the world not thoroughly looked into; and horned animals may yet be found which do not chew the cud.

In like way the most probable — what we call *certain* facts of nature — are open to question notwithstanding their universal acceptance.

The question has been much debated as to what principle underlies Induction; and many answers have been given. Whately says it is resolvable into the belief in the "Uniformity of the laws of Nature," — that this is the major premise, and the particulars observed, the minor; and the whole process syllogistic.

Stuart Mill gives in his adhesion to this proposition as the "fundamental principle or general axiom of Induction"; and yet he says that this principle is the result of Induction, and that it "is one of the last, or at all events, one of those which are latest in attaining philosophical accuracy." But he also says, "Unless it were true, all other inductions would be fallacious." It is hard to see how a thing can be the principle upon which a process depends, and yet be the result of the process; but he is at least right in his assertion that it comes late; and just for this reason it cannot be the conscious basis of a process which goes on long before its discovery.

It is not necessary to speak of other and more elaborate theories to account for this rational process. In the light of what has been said, Induction is just what it seems to be, — a systematic gathering together of facts by the inherent power of personality. The self is by its nature a unifier. It cannot help co-ordinating all facts which are presented, with the purpose of discovering their meaning; that is, the dependence or the relations which subsist between them. When several facts present a common element, since the apperceiving nature of the self is to discover order or law, these agreements lead the self to look for the like element in further instances of the same kind; and thus the expectation rises that further instances of the class will fall under the tentative rule so far formulated. It is not, thus, that there is any original notion touching the uniformity of the laws of nature; for this has yet to get itself recognized, but it is that the self by the law of its nature — the spring of all knowledge — is a diviner of law. To ask why this should be is to ask why man should be as he is — a manifestly idle question. This power is the same as that which at a later stage is called the 'Scientific Imagination.' It projects an order of sequences which has been already noted, into the future as a possible, then a probable, and finally as an established law. And this is all we can ever know of any law of the external world. It cannot be apodictic, or so necessary that its contradictory is unthinkable. It is at last but a fact of experience and can only get such a measure of certainty as its universality may warrant.

It is not possible to make a complete inventory of the necessary facts in this higher plane of our psychical nature. No doubt the sharpest scrutiny must leave behind many which enter into the daily life of personality, just as there are, indisputably, many physical facts which science has not, and perhaps can never wholly bring within the sphere of the understanding. Science is daily enlarging our scope in the physical domain, and philosophy, if we are to make a distinction where none really exists, is giving us clearer views of the higher facts and their attendant laws.

It seems important, however, that we should look at some of these fundamental truths of the hypersensible world, with some degree of attention. First, Space and Time. The notion of space is the necessary logical ground of external reality; we do not come into the world with the notion ready formed, but with a nature such that the notion emerges with absolute certainty upon an acquaintance with the external world, and once in the mind, its contradictory cannot be conceived. It is not necessary to discuss further the question of its genesis. It is enough that it does certainly emerge, and when once in the mind admits of no question. Whether it is itself an objective reality — that is, 'thing' — or whether it is but a necessary form of thought, as held by Leibnitz, urged later by Kant, and finally by Lotze, need not occupy us here. Sure it is, that, we cannot think of object without the spacial notion obtruding itself upon us, and when once in sight, we cannot but admit that it has been all along assumed or presumed in contemplating extension of any sort. All men, in all time and in all places, show themselves to be possessed of the notion, and it is therefore properly called a pre-supposition.

But object must have limits or environment, and it cannot be certainly known until the distinction of the 'thing' from what it is not is clear in consciousness. This requires that the environment of a thing shall be known, before the thing can be defined. The logical environment of object, when all else is removed, is space; and if the object be supposed to dwindle until it shrinks up into zero, the space occupied by it at the beginning still remains, so that objects carry the necessary notion of space out of and beyond them; and, once in the mind, it must remain, though they be removed or thought out of existence. Now space cannot be thought to diminish or move. It is a necessary notion.

But space as environment must be given an objective existence, at least in thought. It must, therefore, itself have environment. What is it? If any limited portion of space be conceived, as a cube, its environment can only be the space without. If it be enlarged, the external space does not retire before it, but passes within. The only limit of space, therefore, is space; that is to say, space is self-limited, or, which is the same thing, unlimited. The unlimited is the non-finite — the Infinite. Thus, the necessary notion of the Infinite emerges in the Self.

Sir William Hamilton and his school contend that we cannot know the Infinite except as a negation, or by 'a thinking away from it.' If this be understood to mean that we cannot construe or limit the unlimited, it is simply a truism. The infinite clearly cannot be shaken together and compressed so as to be got within the compass of the limited, the 'conditioned,' which is the province of the understanding. But Sir William contends earnestly for a higher province of the Self, that of necessary truth, — truth known through the 'Regulative Faculty.' The 'understanding' has not then an exclusive claim to all knowing, else what would become of this 'Regulative Faculty' for which he makes such absolute demand? What would become of the principle of identity, or cause, or the notion of existence? There is a Knowing, — the deepest of all, which lies quite below the plane of the 'Elaborative' faculty, — quite beyond the province of the understanding, — a knowing which is itself the ground of all elaborations of the faculty of relations. The Unconditioned, the Infinite, the Absolute, the Un-caused, the Omniscient, and Omnipotent Pure Being, and all the necessary notions of the Self, are incontestably known somehow; and the Understanding is the instrument through which they are revealed to consciousness, and in attempting to construe them fails not to recognize their inscrutable character. Sir William would no doubt have freely admitted all this, and perhaps he has done it, in effect, in his letter to Calderwood on the subject, in which he says: "When I deny that the Infinite can by us be *known*, I am far from denying that by us it is, must, and ought to be, *believed*." That is quite sufficient, since a necessary belief is the surest of all knowing; but it seems a pity that he did not see the bearing of it more clearly, or seeing, did not make it more explicit. It might have saved the revival of the Sensational School of Philosophy, and the consequent flood of Agnosticism. The same remark applies to Dean Mansel's "Limits of Religious Thought," though it seems odd that any one should fail to see that this is his true meaning.

CHAPTER XVIII.

THE BEARING OF EMPIRICISM ON PERSONALITY.

Intuition of Time. Time the ground of Motion Space, of Mass. Cause conditions Space and Time. Inertia. Self-activity inconceivable in 'things.' Personality the only ground of efficient cause. 'Persistent Force.' Doubt as to the being of 'force' as an entity. Professor Tait quoted. Spencer's effort to find an ultimate Reality. Energy implies Personality. Spencer's position sounder than that of his followers. Quoted. His 'Unknown' known.

TIME is the logical ground of change. There can be no motion, no becoming, out of time. It is therefore a necessary presupposition of motion, as space is of mass. The time-object is an event. Its environment is that which precedes or follows. An event can only be limited by an event, and so all history is but an account of actions, or of things and places as accessory to actions. Space and Time are, as the mathematicians say, incommensurable, they have no common unit. No possible effort of thought can find a passage-way from time over into space, or from space into time.

Time and space, then, taken together, are the presuppositions or ground of all things, — time, the rational condition of motion, — space, that of mass, — motion and mass being the only two factors which physical science has found irresolvable, and into which all physical phenomena can themselves be resolved. But now we find that they each have a psychological basis, and that these necessary presuppositions of the self are the ground of all physical reality.

An object stands related necessarily to its environment, and the ground of the object, as well as that which lies beyond and out of the object, is extension or space. The external space is a reason for the limit which defines the object, and the object is a reason which defines external extension in the direction of the object. Take away either of these, with respect to the other, and that other disappears from thought. They must, then, stand related as cause and effect.

But we have seen that in order to have a notion of a limited portion of space, as of a geometrical cube, we are compelled to give it an environment, and that this environment is itself space. If we do not conceive space as excluding itself, we must give up our notion of a definite portion of pure or geometrical space. There is a reason, therefore, for any possible conception of dimensions or defined extension. But a reason for a thing is its cause; and if one definite portion of space did not support or exclude the space without, it would all rush together into a point and disappear. This seems to be a necessary notion, regarding space as a somewhat having an independent existence external to us, or, as a rational intuition, with only a subjective reality. Space, then, has a ground deeper than itself in the notion of cause.[1]

The case is the same with Time. An event is defined or limited by other events, one before and the other after. Removing these, the event itself must pass out of thought. But when we consider a portion of pure time, as a minute or any fraction of a minute, the limits are still the preceding and succeeding instants, which are reciprocally the reasons one for the other. If the defined instant is not thought of as excluding those on either hand, time can no longer be thought of as enduring, but must collapse into zero and disappear. The notion of causation, therefore, cannot be got rid of, even in time, whether it be regarded as objective or subjective.

But now that we have found causation to be the necessary

[1] See W. T. Harris' "Philosophy in Outline," reprint from *Journal of Speculative Philosophy*.

ground in the world of extension and of duration, that is to say, of all that we know in the external world, we are compelled to allow causality itself to have a ground deeper than itself. A cause to be a cause must have an effect, and an effect to be an effect must have a cause. But these must in some way affect each other. The cause must produce or compel its effect, and the effect must necessarily receive and store up the cause. This is energy, kinetic (moving) or potential (quiescent).

But energy must proceed or go out. This it must do of itself, or it must itself be compelled; that is to say, it must be the recipient of action. In the world of inanimate things, we never think of one thing acting upon another, without being itself first acted upon. A stone would lie just where it is forever, if not disturbed by some external energy. It would never change the state or condition, *i.e.* its molecular condition, unless it were acted upon by moisture, heat light, electricity, or some mode of energy; and when so affected, it converts kinetic energy into potential; which it in turn gives forth again as kinetic; but only when some change takes place in its environment. It is inert; and this principle is the fundamental postulate of mechanics — Newton's first law. It is the principle of Inertia.

It is important to get a clear notion of what is meant by Inertia. It is that property of matter by which change is resisted, with respect to either rest or motion. It is thus always in the opposition, — its voice an eternal ' Nay.' It is the all-pervading recalcitrant factor of external nature; and just for this reason, the conserver of the material Universe. But for inertia, nothing would lie still, and nothing could move, since a breath would move the world, and a breath would stop it when in motion. It is the one necessary condition of matter.

In consequence, then, of this conserver of the material Universe, inert objects have not the power of self-action, but can only be the agents of such energy as may be transferred to them. When we say that a stone is the cause of the breakage of a pane of glass, we know well enough that we do not quite mean it. We know that there must have been a hand and arm to set the stone in motion, if it have a human origin; and when we have traced it back to such a source, we seek no further, but are satisfied that we have found the cause in the true sense.

An action belonging to the class in which one insentient thing acts upon another, is called a secondary or occasional cause. One belonging to the class in which we find the origin of the action to be a person is called an original or an efficient cause. When an action is the result of the working of nature, as when we say, 'The sun is the cause of light and heat,' or 'Matter is the cause of gravitation and inertia,' it clearly belongs to the class of secondary causes. We could think of the sun being the cause of its own action, only upon the precedent thought that it has personality. It is impossible for one to predicate self-action of anything without assuming sensibility as a necessary postulate, — such sensibility, too, as makes it possible for the animated being to put forth a purposive action. This, I think, we are warranted in saying, is the universal conviction of men. And thus it is that in Rhetoric we have the figure of personification in which inanimate objects are, in fancy, invested with life and self-activity. The admission that an inanimate object could act of itself, would utterly overthrow the whole science of mechanics.

It seems clear, then, that the only class of actions which can be properly called causal are original or efficient causes; that is, actions of sentient beings. The self, therefore, or personality is the ground, or presupposition, of cause: and thus we are once more back at the universal and necessary postulate of all certitude, the ego.

There is another point which has an important bearing on

this subject. As we have so often said, the steady trend of all science is to resolve all phenomena of nature into mass and motion, and the one branch of science to which every other looks submissively is mechanics. The leaders in it have been forward to look into the grounds upon which its conclusions in molecular and molar action are based, and have found some remarkable things.

Since the days of Sir Isaac Newton, the march of this branch of science has been steady and brilliant. The one concept about which the whole system has revolved is 'force'; and the philosopher, *par excellence*, of the physical side of nature, has, in our day, staked his whole philosophic fabric on the postulate of 'Persistent Force.'

It is important, therefore, that we should consider this concept for a moment; and it will be interesting to note the present bearing of science in this respect upon the Spencerian philosophy. Professor Tait, in his "Recent Advances in Physical Science," speaking of Force, says: "The notion is suggested to us directly, by the so-called 'muscular sense,' which gives us the feeling of pressure, as when we move a body with our hand or foot. But we must be particularly cautious as to the way in which we treat the evidence of our senses in such matters. Think of Sound and Light, for instance, which, till they affect a special organ of sense, are mere wave motions. The sensation is as different from the cause in such cases as are the bruise and the pain produced by a cudgel or a cricket ball from the mere motion of those portions of matter before impact on a part of the human body. In all likelihood a similar (probably a more sweeping) statement is true of force.

"The definition of force in physical science is implicitly contained in Newton's first Law of Motion, and may thus be given: Force is any cause which alters or tends to alter a body's state of rest, or of uniform motion in a straight line.

"The only difficulty, and it is a serious one, which we feel here, is as to the word 'cause'; for this, amongst material things, usually implies objective existence. Now we have absolutely no proof of the objective existence of force in the sense just explained. In every case in which force is said to act, what is really observed, independent of the muscular sense (whose indications, like those of the sense of touch in matters concerning the temperature of bodies, are apt to be excessively misleading), is either a transference, or a tendency to transference, of what is called energy from one portion of matter to another. Whenever such a transference takes place, there is relative motion of the portions of matter concerned, and the so-called force in any direction is merely the rate of transference of energy per unit of length for displacement in that direction. Force then has not necessarily objective reality, any more than has Velocity or Position. The idea, however, is still a very useful one, as it introduces a term which enables us to abbreviate statements which would otherwise be long and tedious; but as science advances, it is in all probability destined to be relegated to that Limbo which has already received the Crystal spheres of the Planets, and the Four elements, along with Caloric and Phlogiston, the Electric Fluid, and the Odic or Psychic Force."

Under the title 'Mechanics' in the *Britannica* he treats of the question more at large. I quote as follows: "So far we have treated of force as acting on a body without inquiring whence or why; we have referred to the first and second laws of motion only, and have thus seen only one-half of the phenomenon. As soon, however, as we turn to the third law, we find a new light cast on the question. Force is always dual. To *every* action there is *always* an equal and contrary reaction. Thus, the weight we lift, or try to lift, and the massive gate we open, or try to open, both as truly exert force upon our hands as we do upon them. This looking to the other side of the account, as it were, puts matters in a very different aspect.

'Do you mean to tell me,' said a medical man of the old school, 'that if I pull a "subject" by the hand, it will pull me with an equal and opposite force?' When he was convinced of the truth of this statement, he gave up the objectivity of force at once.

"The third law, in modern phraseology, is simply this: *Every action between two bodies is a stress.* When we pull one end of a string, the other end being fixed, we produce what is called *tension* in the string. When we push one end of a beam, of which the other end is fixed, we produce what is called *pressure* throughout the beam. . . . But in the case of the string, the part of the stress which every portion exerts on the adjoining portion is a *pull;* in the case of the beam it is a *push.* And all this distribution of stress, though exerted across every one of the infinitely numerous cross-sections of the string or beam, disappears the moment we let go the end. We can thus, by a touch, call into action at will an infinite number of stresses, and put them out of existence again as easily. This, of itself, is a very strong argument against the supposition that force, in any form, can have objective reality. . . .

"If we inquire carefully into the grounds we have for believing that matter (whatever it may be) has objective existence, we find that by far the most convincing of them is what may be called the 'conservation of matter.' This means that, do what we will, we cannot alter the mass or the quantity of a portion of matter. We may change its form, dimensions, state of aggregation, etc., or (by chemical processes) we may entirely alter its appearance and properties, but its quantity remains unchanged. It is this experimental result which has led, by the aid of the balance, to the immense developments of modern chemistry. If we receive this as evidence of the objective reality of matter, we must allow objective reality in anything else which we find to be conserved *in the same sense* as matter is conserved. Now there is no such thing as negative mass;

mass is, in mathematical language, a signless quantity. Hence the conservation of matter does not contemplate the simultaneous production of equal quantities of positive and negative mass, thus leaving the (algebraic) sum unchanged. But this is the nature of the conservation of momentum and of moment of momentum. The only other known thing in the physical universe, which is conserved in the same sense as matter is conserved, is energy. Hence we naturally consider energy as the other objective reality in the physical universe, and look to it for information as to the true nature of what we call force."

Such are the reasons from the scientific side for doubting the objective reality of force, and for the substantial abandonment of the use of the word in scientific treatises. There is still much to be said in the same direction from the metaphysical side, but let it pass.

Mr. Herbert Spencer in a note to Chapter VI., "First Principles," tells us that he was in some perplexity in the beginning for a satisfactory name for 'the Unconditioned Reality, without beginning or end,' which was to serve as the ultimate in his system. He could not permit himself to use the word 'energy,' since, as he says, "it is impossible to think of 'energy' without something possessing the energy." He expressed to Professor Huxley his 'dissatisfaction with the (then) current expression "Conservation of Force": assigning as a reason, first, that the word "conservation" implies a conserver and an act of conserving; and second, that it does not imply the existence of the force before the particular manifestation of it which is contemplated.' He goes on to say: "I may now add, as a further fault, the tacit assumption that, without some act of conservation, force would disappear. All these implications are at variance with the conception to be conveyed. In place of 'conservation' Professor Huxley suggested *persistence*. This meets most of the objections, and though it may be urged against it that it does not directly imply pre-existence of the force at any time manifested, yet no other word less faulty in this respect can be found. In the absence of a word specially coined for the purpose, it seems the best, and as such I adopt it."

It seems the irony of fate, that after the care Mr. Spencer has taken not to adopt the word 'energy' which would carry with it the notion of an energizer, and to exclude the word 'conservation' lest it should allow of a conserver, that advanced science should lose faith in the reality of 'Persistent Force,' or any sort of force, and insist upon it that the discarded 'energy,' which the philosopher declares cannot be thought of 'without something possessing it,' is after all the one thing which persists! It only shows the impossibility of arriving at the ultimate in the realm of things. But in fairness to Mr. Spencer, it must be said that he has done much to guard himself against the charge of erecting his splendid fabric on nothing as a foundation, even though there be no such thing as 'Persistent Force.' For he has qualified the phrase in a way which gives it an undoubted somewhat as a content. He says, in the chapter in which the phrase is introduced, that 'By the Persistence of Force, we really mean the persistence of some cause which transcends our knowledge and conception.' 'In asserting it we assert an Unconditioned Reality, without beginning or end.'

This definition is thoroughly satisfactory, and no metaphysician or theologian can reasonably ask more : but the trouble is that it has been too much left out of sight, and the common mind has accepted the phrase, 'persistent force,' as the symbol for something dead and unintelligent. Mr. Spencer perhaps never entertained such a notion, — certain it is that he now earnestly repudiates such a construction. He now holds that his ultimate 'Cause which transcends our knowledge and conception,' is an 'Infinite and Eternal Energy.' How he differs affirmatively in this regard from the most rigid theologian, I am unable to see. His words are clear and sweeping: 'I held at the outset, and continue to hold, that the Inscrutable Existence, which science in the last resort is compelled to recognize as unreached by its deepest analysis of matter, motion, thought, and feeling, stands towards our general conception of things in substantially the same relation as does the Creative Power asserted by Theology.' Negatively, he hesitates to apply the appellation 'Person' to this 'Inscrutable Existence,'—this 'Infinite and Eternal Energy,' lest he should limit it. He would not degrade the Unknown Cause of things below personality, but raise it higher. Well, that is, I take it, just what the theologian would do. He simply does not know how, any more than Mr. Spencer,—he does the best he can, however; and herein seems to be the only difference between the two. In the 'Unknown All-Being' we cannot but recognize the Christian's 'Lord of All Power and Might.'

This is a matter of so much moment that the reader will not, I hope, feel impatient if we stop a little longer on it. And the question to be asked is this: Is the reason Mr. Spencer gives for denying personality to that which he does not now hesitate to call energy, really valid? He says that, 'On raising an object from the ground, we are obliged to think of its downward pull as equal and opposite to our upward pull; and although it is impossible to represent these as equal without representing them as like in kind; yet, since their likeness in kind would imply in the object a sensation of muscular tension, which cannot be ascribed to it, we are compelled to admit that force as it exists out of our consciousness, is not force as we know it. Hence the force of which we assert persistence is that Absolute Force of which we are indefinitely conscious as the necessary correlate of the force we know.'

Since, however, Mr. Spencer now freely uses the word 'energy' in a sense, it is to be presumed, he would not have used it thirty years ago, it is fair to assume that he would not now contend that 'force as it exists out of our consciousness is not force as we know it,' and that, if it were to be done over, he would hardly think the distinction so certain or so essential as to be made the ultimate ground of a system of Philosophy. In point of fact, the definition he gives of his fundamental phrase is so much larger than a fair construction of the phrase itself would warrant, that there does not seem to have been any pressing need for it at all; and when it is further taken into account, that it is at best but a factitious phrase, in the beginning unsatisfactory to the philosopher himself, we may be permitted to regret that it was given such emphasis and importance. The ground of such regret is that, while the author of this stupendous system has been at great pains from the beginning, not to shut up the way towards the very highest spiritual (non-material) conception of the Universe, there is nevertheless an atmosphere everywhere pervading it, which admits of, and to the general reader seems inevitably to lead to, the thought of a soulless and dead source and spring of all things. It is to be freely admitted, I repeat, that such a charge cannot in justice lie against Mr. Spencer, though he saw and expressed the danger of such a misconception; and it is to be regretted, that his hostile critics have not dwelt more upon the phase which makes for the truth as they see it, and less upon that which is so obnoxious to them. Any number of passages could be quoted to show the anxiety of Mr. Spencer not to be set down as a materialist, but take the following, from the last chapter of "First Principles": "The liability to misrepresentation is so great, that notwithstanding all evidence to the contrary, there will probably have arisen in not a few minds, the conviction that the solutions which have been given, along with those to be derived from them, are essentially materialistic. . . . Men who have not risen above that vulgar conception which unites with matter the contemptuous epithet 'gross' and 'brute,' may naturally feel dismay at the proposal to reduce the phenomenon of Life, of Mind, and of Society to a level with those which they think so degraded. But whoever remembers that the forms of existence which the uncultivated speak of with so much scorn, are shown by the man of science to be the more marvellous in their attributes the more they are investigated, and are also proved to be in their ultimate natures absolutely incomprehensible — as absolutely incomprehensible as sensation, or the conscious something which perceives it; whoever clearly recognizes this truth, will see that the course proposed does not imply a degradation of the so-called higher, but an elevation of the so-called lower. . . . Being fully convinced that whatever nomenclature is used, the ultimate mystery must remain the same, he will be as ready to formulate all phenomena in terms of Matter, Motion, and Force, as in any other terms; and will rather indeed anticipate, that only in a doctrine which recognizes the Unknown Cause as co-extensive with all orders of phenomena can there be a consistent Religion or a consistent Philosophy."

This is clear and strong, but it will be quite worth while to hear him speak further upon this point; and I am under a very false conviction if the greater part of those who talk either for or against Herbert Spencer are not practical exemplifications of the truth of the first phrase of what follows, speaking in ignorance, or in total disregard, of the statement which it introduces. He says: " Though it is impossible to prevent misrepresentations, especially when the questions involved are of a kind that excite so much animus, yet to guard against them as far as may be, it will be well to make a succinct and emphatic re-statement of the Philosophico-Religious doctrine which pervades the foregoing pages. Over and over again it has been shown in various ways, that the deepest truths we can reach are simply statements of the widest uniformities in our experience of the relations of Matter, Motion, and Force; and that Matter, Motion, and Force are but symbols of the Unknown

Reality. A Power of which the nature remains forever inconceivable, and to which no limits in Time or Space can be imagined, works in us certain effects. These effects have certain likenesses of connection, the most constant of which we class as laws of the highest certainty. The interpretation of all phenomena in terms of Matter, Motion, and Force, is nothing more than the reduction of our complex symbols of thought to the simplest symbols; and when the equation has been brought to its lowest terms the symbols remain symbols still. Hence the reasonings contained in the foregoing pages afford no support to either of the antagonistic hypotheses respecting the ultimate nature of things. Their implications are no more materialistic than they are spiritualistic; and no more spiritualistic than they are materialistic. Any argument which is apparently furnished to either hypothesis, is neutralized by as good an argument furnished to the other. The materialist, seeing it to be a necessary deduction from the law of correlation, that what exists in consciousness under the form of feeling, is transferable into an equivalent of mechanical motion, and by consequence into equivalents of all the other forces which matter exhibits, may consider it therefore demonstrated that the phenomena of consciousness are material phenomena. But the spiritualist, setting out with the same data, may argue with equal cogency, that if the forces displayed by matter are cognizable only under the shape of those equivalent amounts of consciousness which they produce, it is to be inferred that these forces, when existing out of consciousness, are of the same intrinsic nature as when existing in consciousness, and that so is justified the spiritualistic conception of the external world, as consisting of something essentially identical with what we call mind. Manifestly, the establishment of correlation and equivalence between the forces of the outer and the inner worlds, may be used to assimilate either to the other, according as we set out with one or the other term. But he who rightly interprets the doctrine contained in this work, will see that neither of these terms can be taken as ultimate. He will see that though the relation of subject and object renders necessary to us these antithetical conceptions of Spirit and Matter, the one is no less than the other to be regarded as but a sign of the Unknown Reality which underlies both."

These are the final sentences of Mr. Spencer's first work on Philosophy. It is manifestly unjust to say that he is consciously either a materialist or a spiritualist, though since he says that the antithetical concepts of Spirit and Matter are necessary to us, it should seem that he must be both. It ends with an explicit declaration of the certainty of that ultimate Reality of which he has never for a moment lost sight, — a Reality which he nevertheless continues to qualify by the adjective, Unknown. But how unknown? Plainly not in the sense of the un-thought upon, and not in the sense of the non-existent. It is not so unknown as that he could not make a book about It, — not so unknown as not to be a certainty, a Reality, a Cause, a Power, a Force, — to have Persistence, and some way of manifesting itself in the modes of Matter, Motion, and Mind. That this ultimate Reality may be so far known (and do these specializations include all the phenomena of the universe?) and yet remain unknown in many ways, admits of no question; but the same is equally true of the philosopher himself. He is undoubtedly well known to his friends, and to all the world, but is he not, in many respects, especially in his ultimate nature, utterly unknown? Is there anything which, in like regard, is not unknown? How then can he mean that his ultimate Reality is unknown, except in this same ultimate aspect? And if this be the sense in which we are to construe this formidable adjective, what is it but to say that the unknowable is unknown? If this be the attitude of Agnosticism, then we are all Agnostics.

CHAPTER XIX.

FEELING.

Classification. Pain and pleasure. Sensuous Feeling. Herbartian scheme. Intensity and quality in feeling. Cœnesthisia. Esoteric and Exoteric feeling. The one working from within emerges in the understanding; the other built up through the understanding. Practical bearing.

AS was absolutely necessary, in treating of cognition we had constantly to assume, and in some sort discuss, sensation and will. Indeed, sensation has had almost as much consideration as the cognitive power, while in the physiological treatment it was altogether dominant.

Sensation, lying as it does at the threshold of consciousness, is to be considered — if we are to give order of precedence to these ever varying factors — as the foundation of all emotive activity, with volition as the apex.

Separating the sentient element, as far as we may, from the other modes of personality, we can construct a pyramid which shall roughly exhibit the principal phases of feeling, of which sensation in its lowest, unspecialized form must serve as the base, as in the figure on the following page.

What is meant by sensation in this basic sense, will be sufficiently understood from what has gone before. It is, so to speak, the blind response given through the nervous organism to stimuli of whatever nature from without. It is not yet feeling in any true sense, not yet being sufficiently differentiated to be construed in consciousness.

Out of this vital soil, if I may so say, the tree of feeling springs, with its many branches spreading out into twigs and foliage in infinite variety. There is no possibility of knowing what sensation is, and to ask the question is to ask how the question itself can be asked. It is given us, as it is given the protozoa; and we have but to accept it, and be thankful. We can note, however, the varying phases of this fundamental factor of the

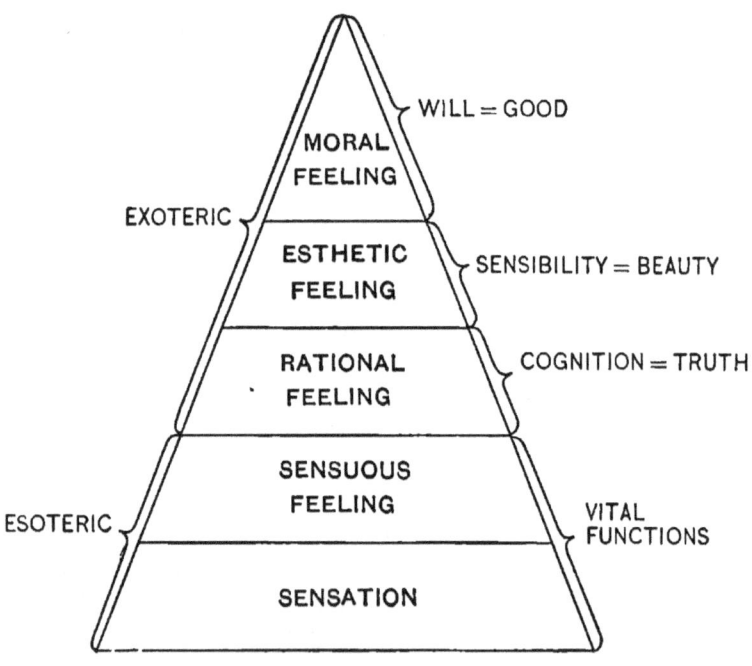

living organism; and we see that in the earliest stages of vitality, the development seems to be carried on through sensation, with resulting movement. At what point in the scale of animate being ideation and purposive movement discover themselves, none can certainly say. It is not, however, until these appear that sensation rises from its elementary form into a complex whole, gradually becoming more and more composite, until at last, infinitely complicated, it becomes what is rightly called feeling.

Sensation, with its primitive features and functions, is never absent in the higher forms of the scale; but still exerts its subconscious energy, just as the foundations of a building ever continue to send upward their reactions to capstone and pinnacle, though out of sight and unthought upon.

What physiology can tell us of sensory and motor actions and reactions throws no light upon what feeling really is. It is, as has been said, rather a matter of being than of knowledge, and gets its meaning through the understanding. Why it is that a bitter taste is disagreeable, and the appeasing of hunger pleasant, admits of no real answer. The physiologist may show that the one affects favorably, and the other unfavorably, the development of tissue, and so hinders or helps the body with respect to its harmonious action; — but while this shows that the one effect is hurtful to a certain end, and the other helpful, it throws no light upon why pain accompanies the one, and pleasure the other.

It may be well, however, to look at the question of pleasure and pain a moment, from a teleological point of view, *i.e.* as to the ends they subserve.

The world is full of people who are constantly declaiming against the fact that there is any such thing as pain at all; and in these last days, certain sciolists, taking this fact as a text, talk wretched nonsense, to speak of it in the mildest terms, about 'a bad God' for allowing it. They do not seem to know that if there were no pain, there could be no pleasure, — no satisfaction of any kind, since pleasure could then have no limitations, and would fall out of recognition. That would be bad enough, but would only be the beginning of the disaster: there could be no sensation of any sort, and no knowledge. Man would be reduced to a senseless mass of inert matter, — if even this could be the end of it.

Pleasure and pain are not 'things,' but feelings, utterly incommunicable. A thought — any kind of knowledge proper (for feeling, though by its nature inseparable numerically from thought, is not thought, and must be translated into thought before it can become knowledge) — may be communicated or shared with another; not so feeling. If pleasure were not so transferred to the cognitive mode of the self, it would be, for us, as though it were not. But we have seen that any concept, to be known, must be negated, and the negative of pleasure is pain; so that if pain were impossible, satisfaction would be also. One could not maintain one's existence without this much maligned element of our nature. In some careless moment, one would burn up, without knowing what was going on; starve, not knowing oneself to be hungry; and so throughout the round of life. Pleasures are not ends in any sense to be pursued, nor is suffering a calamity, *per se*, to be avoided. They are accessory to ends which are higher than pleasure; and are warnings against calamity more dreadful than pain in its direst form.

That pleasure and pain are only known in relation is shown by the fact that a diminished pleasure is a pain, and a diminished pain a thoroughly recognizable satisfaction. The contrast may vary through many stages. That which gives one pleasure to-day may be the source of pain to-morrow. It is the relation which determines the degree. A sum of money lost, or gained, produces a vastly different feeling at one time, and another. The indulgence of an appetite may be highly satisfactory at night, while horrors may be the result in the morning.

Sensuous Feeling, the lowest form of sentient energy which can be rightly called feeling at all, embraces all specialized sensations as to place and kind, — all agreeable and disagreeable tastes and smells; all the recognized affections of the self through the special and general senses,— in short, all bodily affections which give rise to satisfaction or infelicity.

As we have seen, feeling to be known must be translated into thought; and so psychologists of the Herbartian school work out a very complete scheme to account for the play of the emotive nature in what they call 'furthering' and 'arresting' concepts. The state of consciousness is like the rapid changes which we often see in the clouds under the action of cross currents in the air. New concepts are constantly rising to contest the place of the old. These do not yield without resistance; the new and the old meet with every variety in force and direction, the main divisions supported by many auxiliary forces on either side; and so the contest, sometimes in comparative peace and gentleness, and at others, with rapid and violent movement, but never ending, goes on through life. Assuming that there is a measurably constant quality of psychical energy when no arresting concepts are present, the flow is regular, and there is a general feeling of satisfaction; when a new concept acts *with* the current, a positive pleasure; but arrest gives rise to a partial stoppage, which is depressing or painful. A feeling is thus the consciousness of the furthering, or arrest, of the flow of thought,— pleasurable if it be with, painful if it be against, this movement.

This still does not really tell us anything of what feeling is. It but gives us the thought-basis of feeling, or the manifestations in thought, as the feeling may come from the lower or higher states of the self; but these are not feeling itself, any more than the action of sensory and motor nerves are sensation.

It is hardly necessary to enter, in any detail, upon the question as to how localization of sensations by which they gain character and rise into the light of consciousness, take place. The writers on the side of scientific psychology, such as Herbart, Bain, Wundt, and Spencer, have gone over the ground with infinite pains, and with such completeness of detail, as would prove wearisome to the general reader, even in outline, with no corresponding profit.

We pass on to a summary of a more general character. The gustatory senses, taste and smell, the lowest in the scale for scope and variety, are chiefly charged with the duty of provisioning the physical organism. They afford the basic motives which, taken with the pains of appetite, induce original effort. Food is made pleasant to the palate, or we could not maintain our existence. If a resistance had to be overcome at every morsel which enters the mouth, or if it were a matter of indifference, the child would have to be forced to take nourishment, or rather, could not be made to take it at all, and would die at the start. It would be no better with the adult, if any there could ever be. The teleological value then of pleasure in the reception of food is plainly manifest. It is perhaps not too much to say, that many people seem to be satisfied that they have found the end and object of life, at this mere threshold of existence, and make ends of what are simply means.

It is to be noted, that in the pleasures of the palate there is the same ascending scale which we have found everywhere, and which will continue to meet us as we go on. In the beginning of life only the simplest flavors and savors are pleasant. The child resists vigorously nearly all condiments, such as pepper, mustard, and the higher spices which the cultivated palate finds quite necessary to remove insipidity; so that finally the highest satisfaction of the gustatory sense lies clearly in the domain which at the earlier stage would be pronounced decidedly painful. In the lower plane, the stimuli are gentle, becoming more intense as we ascend. This difference in gradation is called 'quantity' or 'intensity.'

All pleasures and pains have this quantitative characteristic. Any sensation raised beyond a certain point, differing for different persons, passes from pleasure into pain, or the reverse.

These limits cannot be definitely fixed, and are not constant. In sound, if the attention has to be strained, or if there remain vagueness and uncertainty, the effect is unpleasant; if raised beyond a certain degree of loudness, it is also painful. Between the two lies agreeableness, reaching the maximum at the point where there is an easy flow without stress upon the organism. With light it is the same; and so with the tactual or any other sense.

But this point of maximum agreeableness does not remain stationary in the same organism for any considerable time. The same unvarying sensation long continued, though altogether agreeable at first, becomes monotonous, and even painful. This characteristic is called the 'duration.' Sensation must not be too short, for then there is a baffled feeling; it must not be too long, or weariness results. Continued uniformity is impossible. Even an acute sensation is greatly abated by long continuance, so that at last it almost passes out of consciousness. Thus there is a constant adjustment of the organism to defeat continuance of either pleasure or pain.

There is also what is called 'quality' in feeling. This refers to the kinds of stimuli whence the feeling arises. There is a marked difference between a pleasure due to sensation through touch, and one granting the degree of satisfaction the same, through sight. Sir William Hamilton, following Kant, maintains that there is an inverse ratio between feeling and knowledge, that in the senses beginning with touch, in which sensation is at a maximum, there is a marked diminution up to sight, in which the sensuous element is no longer distinguishable; while, on the other hand, cognition is at a minimum in touch, and sensation at a maximum. So it is, in general, in the relation between feeling and thought. In the higher stages of feeling, the sensuous factor almost, or entirely, drops out, so far as discoverable in consciousness. A scale arranged on this basis would give us differences in 'quality.'

Mention has already been made of what is called the 'sensus communis,' or 'common feeling,' falling under the general head of *cœnesthesia*. A word or two only will be necessary here. The feeling of buoyancy, or depression, plays a very conspicuous part in every one's life; and it is, therefore, of great practical importance. It has been called the massive or voluminous sense; perhaps because one has such a large opinion of oneself when everything seems to go one's own way. The flow of the intellectual and emotional life is in full sweep, the countenance cheerful; and, if one be young, or fail to exercise a dignified restraint if old, gladness breaks out in laughter and frolics. It is the joy of living. The reverse state, known generally as the 'dumps,' or 'blues,' presents a melancholy contrast. The world is out of joint, last friend gone, the bottom fallen out, etc., and all from no other reason than the depression of general tone. If, however, these results are due, as they may be, to known causes, they are removed from this classification.

The sensations due to specific touch also belong to this class of Sensuous Feeling, such as soft, smooth, rasping, etc. Many people show marked peculiarities connected with this lower form of feeling. Some organisms are very disagreeably affected by the touch of velvet, the fuzz of a peach, and many like substances. Also by certain perfumes and flavors. The effects are sometimes uncontrollable and violent. All that can be said is, that these effects seem to be due to idiosyncrasies of organism.

There is a marked difference between the lower group of the senses, — touch, taste, smell, as well as all the newly discovered senses, — and the higher group, hearing and sight. Sensation excited by stimulating any one of the lower group is purely subjective, and has localization within the body; on the other hand, sensation in the higher group has an objective character, the localization being external to the body. For this reason the lower class of sensations may be called 'Esoteric,' and the higher class 'Exoteric.' In the esoteric class the feeling excited carries with it the consciousness of the mechanism, or instrument of sensation; in the exoteric, on the contrary, the feeling is quite independent of any consciousness of bodily affection. Thus, in the flavor of an orange, I am conscious of the taste in the mouth; but with the color of a violet, consciousness declares the color to be, not in the eye, but in the flower itself.

This difference marks an important distinction between the entire domain of 'Sensuous Feeling,' and that immense range above, which we have called the 'Rational,' 'Esthetic,' and Moral Feelings. In the one case we have feeling, simply recognized as existing; in the other, there is no proper feeling at all, until *perception* of that which is without produces a reaction in the sensibilities: that is, in sensuousness, the sense element is primary and the intellectual element secondary: in the higher range, the thought element is primary, and the emotional element secondary.

This has an important practical bearing, since it thus appears that, in the psychical factor of the personality, there are two well-marked domains of feeling; the one essentially fundamental, as pertaining and ministering to the animal mechanism; the other, acquired through self-activity, is characteristic of the psychical factor of the personality. Feeling, from the lower plane, is simply forced upon the personality, and suffers very little modification; feeling, from the higher domain, is secondary, being derived through the rational efforts of the self, and depending upon it for its infinitely various shades of refinement and intensity. The one comes from below, and is common to man and the lower animals; the other comes from above, and is the peculiar heritage of man alone. The true work of a rational creature is, negatively, to regulate and hold himself in check with respect to esoteric feeling; positively, to expand, refine, and elevate his nature in the domain of exoteric sensibility. Yielding to the one, he becomes a slave and retrogrades towards mere animality; cultivating the other, he rises into freedom and lives upon the heights.

The three main divisions of that class of Feeling which I have ventured to call Exoteric are specialized, each, by one of the three fundamental modes of the Self: thus Rational Feeling finds its characteristic in Cognition, with Truth as its content; Esthetic Feeling in Sensation, with the Beautiful as its content; while Moral Feeling is dependent on the Will, and has the Good as its content. Now as there is an underlying unity in the three several notes which give character, respectively, to these different forms of feeling, so we cannot but think there must be unity, also, in their content; the ultimate ground of all being the one Infinite Personality.

The primitive desires of our nature are founded essentially in Esoteric Feeling. They are blind, and ray out in every direction, seeking satisfaction. In their original activities they are, in a moral point of view, neither good nor bad, — have no moral quality whatever. But experience soon teaches us that the unbridled indulgence of these original propulsions results in pain and injury, and we are compelled to practise self-restraint. This is the beginning of personal development, — the 'Thou shalt not' of personality.

But there is another class of desires, the Exoteric, built up through the objective senses, by means of cognition and will. Such desires are not absolutely independent of esoteric feeling, — there never can be any possible emotive activity that is; but they have an objective character, and come into us, so to speak, from without, by our own conscious activity, instead of being beforehand with consciousness, as is the case with the esoteric class. Such are the desires for power, wealth, learning, fame, and all the ends for which men strive, and to which they attribute value, from a discovery of their meaning. The character and intensity of these desires will depend upon the values discovered to be in them, or rightly or wrongly assigned them by the power of the understanding, and the determination to pursue or turn away from them. In the beginning, the emotive element with respect to them is feeble, and it is in man's power to keep it so; but it is also in his power to increase it, not directly, it is true, but indirectly, by dwelling upon, and establishing more firmly in the mind, the worth and importance to him of the particular end which the desire subserves. The effect of environment in this work of building up the emotive phase of personality is obvious. Parents, teachers, societies, are constantly emphasizing the several ends which seem to them of paramount importance; and the child, the youth, the man, has borne in upon him true or false values, which being assented to or repelled by him, develop his emotive nature, and so the personality, for better or for worse. This of course involves the regulation and control, as well, of the esoteric class of desires. They, by proper handling, may be greatly refined and elevated; but so long as the vital tides are strong, they continue to move us vigorously towards their blind ends; and so, we find the best of men, in moments of weakness or under abnormal temptations, thrown off their balance, and yielding to what is in glaring conflict with their higher natures.

Passion is a temporary release, and abnormal stimulation of the lower desires.

In the nature of the case, we cannot know what ultimate Truth is, since we should be compelled, in a final analysis, to explain it by means of itself. When we say, therefore, that the ground of Rational Feeling is the True, we have reached the utmost limit in the way of explanation, and must turn back and content ourselves with the divers forms under which it presents itself to our powers of apprehension. By virtue of our human personality we do apprehend it, with more or less readiness and certainty, and, so apprehending, love it, with varying degrees of intensity, depending primarily on our native susceptibility, but perhaps more largely in the degree and quality of the development in us of what we have called exoteric sensibility. Rational Feeling is thus an immediate response of the self to discovered truth, just as the sense of touch is the response to an external stimulus.

But the important distinction must be borne in mind that a sensation coming to us through the stimulation of any one of these esoteric senses is immediate, and the function of the understanding is to construe it; while in the higher feeling no sensation comes or can come to us until it has first passed through the scrutiny of the understanding. The self-developed sensibilities then respond according to the character and degree of such development, and independently of volitional control.

The question as to what is true is purely a matter of the understanding, and its conclusions vary through a large range for the same person at different stages of enlightenment, and through a larger range for different persons, under varying circumstances of time, place, and education. The understanding, however, having once pronounced, the developed sensibilities respond at once upon the presentation of the proper thought stimuli; and thus it is that the same subject or thing affects different persons, or the same person at different times, so variously.

CHAPTER XX.

FEELING (*continued*).

Rational feeling. Esthetic feeling. Beauty. Periodic motion. Music. Vision. Illusions. Berkeley's 'Theory of Vision.' Knowledge through vision. Cheselden's Case. Other cases. Problems mentioned.

'RATIONAL Feeling' is that satisfaction which arises in us from the discovery of the reasonableness of things, — a sense of harmony among varying phenomena; and, on the other hand, the annoyance and perturbation resulting from a baffled effort to find the law, which we somehow know obtains, though not yet recognized. The sensation is one of effort, resulting, if effective, in an exaltation, — a feeling of victory and triumph : if unsuccessful, in a sense of impotence and disappointment. The self, by reason of an inborn energy, tries always to solve difficulties, and find the key to phenomena; if the effort be baffled, there results a sense of failure which shows itself in fretfulness. When once an inquiry is raised, and there is sufficient interest engaged, the self is never satisfied until the relation between phenomena is clear. Confusion is a source of annoyance, and is resisted. When the clew seems at hand, we are pervaded by a pleasurable eagerness,— a feeling which every one has experienced in solving a mathematical problem, or working out a puzzle. So, also, we often feel surprise and bewilderment in the prosecution of an inquiry.

The ground of the emotion in all this vast domain is intellectual; and the end toward which it moves us is truth. It has been called by psychologists, 'Logical Feeling,' and also 'Formal

Feeling.' It is especially dominant in scientific inquiries, resulting from the exercise of the 'Scientific Imagination.'

In 'Esthetic Feeling' the ground is Sensation; but it is Sensation projected upon reality without, in space- and time-forms. It is essentially a somewhat felt—Rational Feeling being a somewhat understood. We *perceive* the beautiful in form and tone, through Esthetic phenomena, — we *conceive* the True through the relation of things and events.

By the logical powers we comprehend the meaning of the external world, and find a response in our emotional nature which loves the True :—without stopping to construe or ask why, we see and hear in the world without what is pleasing to eye and ear. One is a pleasing thought; the other is a pleasing sensation. Thus, in a general way, the Beautiful is the object of Esthetic Feeling; while Rational Feeling finds its satisfaction in the True.

Now it is not to be seriously doubted that Truth and Beauty are bound together by an indissoluble tie, and that, to an intelligence sufficiently exalted, the true would always appear beautiful, and the beautiful true. As it is, we can discover something of this nexus.

We have already seen, in speaking of the senses from the physiological point of view, that vibration is the physical basis of the sensation of sound. The sense of hearing occupies an especially important place as opening the way to the whole rationale of science, in the all-embracing doctrine of undulations. The motions in sound phenomena are of such a sluggish character, that they can be made apparent to the eye and touch, and yet it can hardly be said that vibrations are known at all to the hearing. Whenever pulsations reach a sufficient frequency to produce a continuous sound or tone, the vibratory character is swallowed up in, or rather, is, for the feeling, the tone itself.

It is through this doorway of sound that we get a look into

the rhythmic wonders of creation, and have at the same time an exemplification of Rational and Esthetic Feeling. It is here that we catch the first glimpse of that magic and mystery of motion which has come to be the scientific postulate of all possible phenomena, and which is the only explanation science has to give for sound, light, heat, electricity, — the heavens and the earth. We look not with the eye, but the imagination, and see periodic motion — motion back and forth — in circles and ellipses and straight lines, — in parabolas and hyperbolas, — in every conceivable variety of the conic curves, — all keeping time in a sort of a rhythmic dance; and this is sound. What possible likeness is there between the feeling excited by a flute note, and the rapid swaying in and out, from one oval shape to another, of the embrasure into which the air is breathed, and the witches' dance of air particles caused within the tube! What possible nexus in thought is there between the delicate tints of the violet, and the million times more rapid scurry to and fro of the luminiferous ether! This is strain enough upon one's fancy, but when we are required to see in the solid and immovable door-knob, nothing but an infinite play of vortical motion, and admit the deadest of all dead matter to be but motionless motion, old-fashioned common sense feels abashed. When we have got this far it seems too late for one to cast scorn upon the famous inventor, who, it is said, expects to propel an ocean steamer across the Atlantic, with the movements of a fiddle bow!

But, not to quite lose ourselves in this mazy world, if we like not to give in to the demands of science, we shall find great difficulty to discover a place where we can halt, and say, 'thus far and no farther.' We shall be beaten back until all explanation must be abandoned, and find ourselves in that other world of magic, so familiar to the nursery. We are compelled to fall in and move with this 'fleeting show,' finding it after all the same old commonplace world; only convinced that we do not know as much as we thought, before we tried to look beneath the surface of things.

One thing seems plain through all this wonderland; and it is that periodic movement lies at the bottom of it. This is the rational basis of what we call rhythm. This rhythm is, however, infinite in variety, so that perfect uniformity, perfect agreement, is never to be found.

If I may so express it, nothing in the Universe exactly fits. It is a trite saying that no two shells on the seashore are alike, and no two leaves upon the trees; much less are any two faces — any two voices — any two thoughts, — alike. Likeness is no more a law of the Universe than difference; harmony no more discovers itself than dissonance; except that we are compelled to recognize the one as positive and the other negative. In the movements of the heavenly bodies there seem to be no two exactly commensurate. The period of the revolution of the earth upon its axis is no exact part of the periodic time of the earth around the sun; nor is this period exactly measured by any other of the celestial revolutions. In music we have the same thing. The notes of the diatonic scale do not present perfect agreement in intervals. There is 'the little rift' — always a little too much, or not quite enough. There is no 'dead point' in the mechanism of the All-Father. He is the 'One in the Many,' and the 'Many in One'; and whatever comes from His Hand defies the ultimate scrutiny of man.

Writers generally are pretty well agreed in ranking light above hearing, and no doubt they are right, considered from a utilitarian point of view; but it may be seriously questioned from the sentient and emotional standpoint. Remove speech and music from the world, and with them would go perhaps the larger part of esthetic enjoyment. The range of the auditory sense is far greater than that of sight. The octave above octave finds nothing like it in any other sense. We do not know, and cannot conceive, what the spectacular effect would be if light built itself up, tier on tier, like music; though it is possible that in higher orders of being such color-effects may actually obtain. But certain it is, that the heart and imagination, through this wonderful fact in sound, are wrought up to diviner heights than can be produced through the eye. Not only has sound this marked advantage over sight in altitude, but its advantage in expanse is almost as remarkable. Sight takes in only one-half the circumference about us, and that imperfectly, but hearing embraces the whole circle. The eye must be directed by muscular movement, — the ear is always ready to respond to whatever is within the radius of its power.

The pleasure given one through the ear is largely augmented by cultivation. There is the same ascent from the simple to the complex which we have noted in the senses already considered. At first the simple unison of sounds is a delight; and the untutored ear enjoys harmonies which are open or far apart. Two persons singing together, one an octave above the other, give the untaught ear satisfaction, but not for long; then the fifths and the thirds. The 'chord of the tonic' is a joy; but, after a sufficient cultivation, a closer harmony is demanded, until at last, positive discords greatly heighten the effect. So, again, in melody, the succession must be simple, and the recurrence of passages frequent, to engage the interest of the multitude. If there is no 'tune,' the attention flags, and, instead of pleasure, positive distress results. Hence it is that the populace are always calling for simple airs, in which the rhythm is so marked as to carry head and feet with the movement; while the sonatas of Beethoven and the massive harmonies of Wagner are a mystification and a torture; but for the cultured ear, tune is no necessity, and the divine productions of the great masters produce a 'joy of elevated thought.' The simple child-like satisfaction — more the result of an exuberance ready to burst forth at any sign — gives place to a rational elevation, which revels in minor effects, and causes the emotional nature to stand on tiptoe, striving vainly to look into the infinite and mystical, with a 'joy which is akin to woe.' Feeling thus mounts up, as if to seek another world, and pleasure and pain strike hands.

Sensation in Vision is extremely complex. Undoubtedly light — marvellously rich in color components — and its negation, darkness, with its infinite gradations of shade and shadow, furnish the material out of which the self evolves beauty of form and symmetry, with all the charming effects in color. How, is quite another question, and raises many points by no means settled. There is difficulty enough in determining how the eye acquaints us with the mere facts of the external world, without the farther inquiry as to how the emotive elements of grace and beauty emerge. It is not to be seriously questioned, — no system of Idealism will ever change the conviction that there is an objective reality which answers to the feeling of truth and beauty in the mind. As Lotze says, "There is an inherent order in things: the forms which they lead us to realize or to rejoice in as manifested in nature, are modes of relation of the manifold into the joy of which we are able to enter." He carries the thought further: "Just as there is no sense-perception without its share of feeling, so, too, the notion of a relationship never rises within us without our testing the special degree of pleasure or of pain which this relationship must confer on the two things between which it exists. We never notice identity without at least a faint recollection of the blessedness of peace, or see contrast without a glimpse, sometimes of the hatefulness of enmity, sometimes of the enjoyment that springs from the mutual contemplation of opposites; we cannot discern equipoise, symmetry, rigidity of contour, without, as we gaze, being stirred by manifold pain and pleasure of secure repose, of bondage under fixed laws, or of limitation and confinement. The world becomes alive to us through this power to see in forms the joy and sorrow of existence that they hide;

and there is no shape so coy that our fancy cannot sympathetically enter it."

How far this is fancy, how far sober reality, we need not stop to inquire; but it is easy to carry the subjective element to such an extreme as to underrate or even lose sight of the Infinite Personality which gives reality to Nature. It has not been the design of science, but the indirect result of our better knowledge of the mechanism of nature that has had the effect of driving out of the hearts of too many people that lively sympathy which men in the earlier ages of the world manifested for her power and beauty.

> "The world is too much with us: late and soon,
> Getting and spending, we lay waste our powers:
> Little we see in Nature that is ours;
> We have given our hearts away, a sordid boon!
> This sea that bares her bosom to the moon;
> The winds that will be howling at all hours,
> And are up-gathered now like sleeping flowers;
> For this, for everything, we are out of tune;
> It moves us not. — Great God! I'd rather be
> A Pagan suckled in a creed out worn:
> So might I, standing on this pleasant lea,
> Have glimpses that would make me less forlorn;
> Have sight of Proteus rising from the sea;
> Or hear old Triton blow his wreathed horn."

Sight is the most accurate and definite of all the senses, and yet the information given us by the eye comes to us through more apparently incompatible data than in any other sense. In the first place, the image on the retina reverses everything, making up down, and right left. This at one time was a puzzle, though it has now pretty well passed out of discussion, since we have come to reflect that we do not see the image at all in consciousness, and would never have known of the inversion except for the investigations of the physiologists. As we have seen, there need be no possible *likenesses* in the molecular reactions of the brain-cells to actual objects in nature, and if there were, nothing would be explained.

Again, we have that whole class of what we call illusions in perspective, — illusions, however, without which we could really see nothing in true relation. But, truth to say, they ought to be sufficiently confounding to the Gradgrinds, who insist upon 'facts, nothing but facts.' Let one look at the setting sun, when broken clouds are interposed, and see the streaks of light raying out, fan-shaped, sometimes through a whole semi-circumference. The eye tells us that they are inclined to each other in every possible angle, and yet they are really parallel; and so, in one degree or another, it is with the whole range of vision. These illusions, everywhere existing unobserved in nature, may be made apparent by a little artifice. Lines exactly parallel may be easily made to look inclined, or curved, by means of auxiliary lines; and curved lines made to appear parallel by the position of the eye. So also straight lines may look broken; angles, larger or smaller than they really are.

As an illustration of how the judgment may be deceived in vision by insignificant means, take this figure.

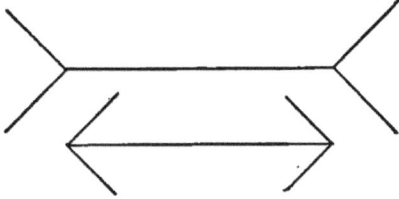

The horizontal lines are of equal length, but they do not appear so.

In the four following figures, in which the principal lines in each form the same-sized square, the eye pronounces unreliable judgments.

It thus appears that our sensations in vision are immensely

composite, and full of aberrancies; but, doubtless, if it were not so, effective vision would be impossible; and we may be thankful that those inflexible souls who would reduce the uni-

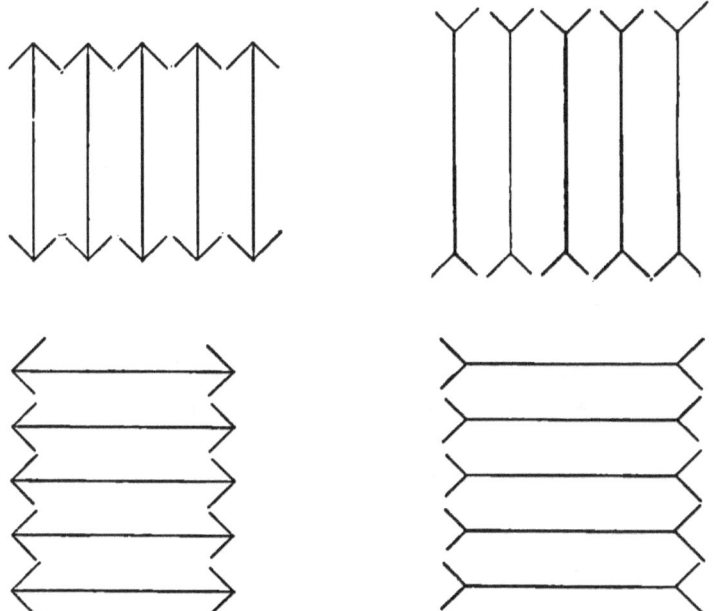

verse to square and plummet are restricted to an exercise of their recalcitrant predilections without the power to set the world right after their purblind fancy.

CHAPTER XXI.

FEELING (*continued*).

Art. Ideal element. Sculpture. Painting. Music. Architecture. Poetry. Evolution of ethical feeling. The good. Ethical treatment reserved to later stage.

IN treating of the Imagination we saw that the Art-world is primarily indebted to that original and creative power for its existence; but it is scarcely less dependent upon feeling. The Imagination is a form of cognition, and as such is cold and powerless. All warmth and intensity must come from the emotional element of our nature, and thought would have no value did it not reach down and quicken the pulsations of the heart. If one could, in thought, penetrate and construe everything, but had no interest in the result, either pleasurable or painful, manifestly nothing would have value, and there would be no inducement for the self — perhaps no power — to prefer one thing to another, or to distinguish itself from aught else: indeed, the very hypothesis is shown to be inadmissible by the attempt to entertain it.

It is thus apparent how large a part the emotional nature of an artist or a poet must play in any effective work. It is not always easy to point out the sentient element which has moved the imagination in the production of a particular work, nor wherein lurks the subtle effect in the work itself; but it must have left its impress behind, or it will not move the heart or fix the attention of another. Hence it is that the artist and the poet must put their own best emotions in their work, or they will fail to move their fellow-men. They must be able to catch and fix the subtle power and pathos which are ever appealing to us for recognition in the processes of nature, and in human conduct, and so help others whose sensibilities are less acute, or whose training is deficient, to feel the truth and beauty of the world-work spread out before us by the Ineffable Artist.

It is not that the artist is to simply reproduce the truth, or the beauty, which he sees in the material world, or in human conduct; for, manifestly, that would help nobody to see or feel, that would leave one still to discern for oneself; but the artist must present the subject of his work with his own personal factor through and through it, if he would give it value. This is what is called 'idealizing'; this is the part which the imagination, the original and creative factor, plays. It is not that one is expected to admire, or wonder at the power and cleverness of the artist, but to hear or see his work with an elevation of heart and soul, looking through the work itself, and the artist's treatment, to celestial wonders beyond. Thus mere imitation is not Art; it is the base and counterfeit in Art. And yet, the personal element must not be obtruded. It must be the 'True' and the 'Beautiful' that one feels, not the artist; the personal element is in the handling by which one is made to feel what would otherwise escape one. There is time and opportunity enough to admire the artist, but it is after the pleasure his art has called forth has had free scope.

The artist cannot invent his truth, nor his beauty; he must take them from the 'pattern in the mount,' and so it is that he must study the form and movement of the actual; but his material in hand, it is for him to produce new effects, and touch the heart by presenting the pleasing and pathetic under new conditions, and with personal emphasis. It is herein that man shows his superiority over nature; that is, nature in its mechanical aspect. He finds himself no longer bound, but begins himself to devise and create. Hence it is that in true art there must always be the real and the ideal; but the real is the work of the bondsman who must do his task; the ideal is the work of the freeman who but uses the drudge to supply him with the material by which he bodies forth thoughts before unthought, and songs till then unsung. An artist must be a realist, but if that be all, his work is only that of a faithful and intelligent copyist; it is as an idealist that the freedom and power of genius opens to us the world of unspeakable things.

It is just for this reason that the sculptor is so restricted in his art. He works in all three of the spatial dimensions, and so, in space-form, his work must be literally and exactly real. Happily for him he cannot carry the realistic element further. He cannot produce flesh tints, expressions of the eye, nor any of the endless color effects. If he could, he would make monsters; we could not endure for long a dead, false thing so entirely like a human being or any real animal as to actually deceive us. We hate what seems to set itself up for true, and is not; but we only begin to hate it, when it begins to so simulate what is real and we feel ourselves in danger of mistaking the false for the true. There is no danger of mistaking the Venus of Milo or the Greek Slave for real women, and so we greet them with a glow of pure emotion; but if we, for the moment, could not distinguish them from living forms, we should turn our eyes away in confusion. It is the non-realistic, or preter-realistic factor remaining to the sculptor that makes his art possible.

In painting, the artist has but two dimensions possible to him; and thus he has, in the start, a great advantage over the plastic arts in the necessity for the ideal creation of the third dimension. It is just by virtue of this restriction — this forced exclusion from a fatal realism — that painting gets its comparative freedom from the trammels which encompass sculpture, and finds a world of ideal possibilities opened to it. There is, however, an element of realism left it, which is its snare, — the power to delineate the texture and tint of the human form, and the expressions of human emotion; and it must be confessed that artists have used their power in this regard to the serious hurt of their art. But this apart, the high idealism of the painter, founded on a true, pure realism, has opened up to us a world of transcendental beauty and excellence.

The poet gains greater freedom still, because he has no trammels of space whatever; and so is released from all possible realism of external form and color. His is the ideal world indeed; but he, too, has his snare in a subjective realism gathered up in words. By means of these he causes the actual world of things and events to pass before the mind and touch the heart. He, too, has often faulted by presenting a gross and impure realism in seductive guise; and in so far has debased his muse.

In music the possibility of realism is almost wholly cut off. The musician, however much he may try, cannot cause the real world of object or action to pass before the mind. In so-called descriptive music there is never more than a faint suggestion of a real action, and none whatever of objects. The sounds of nature are indeed sometimes imitated, in a way, by the composer; such as the song of birds, the rush and roar of the elements, the clank and jar of machinery; but it is a hazardous domain, and if not carefully handled results in the ridiculous and vulgar. In the hands of the great composers admirable effects are thus produced, but such ventures at imitations are always highly idealized, and helped out by suggestions from word and action. The 'anvil chorus' in 'Trovatore' is a success, but nobody would ever have known what it really meant except for its setting; and so in the 'hailstone chorus' of the 'Israel in Egypt' there is only an appropriateness in the music, and in no sense an imitation of the fact in nature; while in such works as the 'Pastorale' the likenesses are purely fanciful.

Architecture is even more tied down to the real than sculpture, since the unideal condition of use enters in such large degree; but it has, on the other hand, a sweep and freedom denied the sculptor, in that it need not be imitative, and may mount up in power and massiveness to the stupendous and sublime.

There is, then, a right place for realism in art, and there could be no art without it; but it is not the factor which gives value to the work of genius, and elevates and refines our manhood. It is the region of determinism, in itself, inflexible and expressionless. It is not until the free creative power of personality breathes into it the living spirit, that truth and beauty spring forth to charm and elevate. It is thus we see how transcendent the psychical factor of personality is, as compared with the stark and rigid facts of mechanical phenomena.

The five grand divisions of art, as commonly reckoned, — Architecture, Sculpture, Painting, Poetry, and Music, — fall into two distinct groups with respect to space and time: the first three being dependent upon space-form, and the last two upon time-form. The first group — Architecture, Sculpture, and Painting — having definite spatial limitations, are utterly immobile and rigid: the second group — Poetry and Music — untrammeled by place-restrictions, are free and flowing, — finding their very life in rhythmic movement and sequence. This emancipation from the bondage of localization raises these two rhythmic arts far above the space-clogged group; but poetry is entitled to a pre-eminence exclusively its own, in that it compels the universal round of feeling to its uses, finding itself equally at home in all three of the great domains, rational, esthetic, and moral; while all the others are almost wholly restricted to the one esthetic mode. It is not to be denied that the truest work of the poetic art is when the esthetic factor is dominant; but there is much high, true poetry founded chiefly in the sense of the true, as well as much that has the feeling of holiness and purity for its characteristic motive. The graphic and plastic arts do unquestionably evoke feelings, both of the true and the good; but such feeling is secondary in character, arising through perception which is primarily esthetic.

Poetry has still another prerogative which sets it apart from and above all the other arts; the catholicity of its content. It uses freely, and as of right, the whole round of material wrought upon by all the other arts; with a world of matter still beyond exclusively its own. Architecture, it is true, makes use of sculpture and painting to a limited extent, and in an auxiliary way; and music borrows the help of poetry at times; but the poetic muse seizes upon, and passes through her alembic, all that appeals to human consciousness in nature and in art. She claims as her own a share of musical effect in her numbers, — revels in the 'long-drawn aisle, and fretted vault,' — not only gazes upon the graver's finished work, but listens to the click of his chisel, and peers in upon the flood of high emotion which fills him with joy, while he compels the marble to show forth the image in his soul. To the poet everything is laid bare, the conflict of motives, — the sweep and rush of passion, — the pathos and joy, hate and despair, sweetness and rest, on earth and in the far beyond. The poet is only limited by his own soul-power, and his genius of expression.

We saw that Truth is disclosed to us through the understanding, and, finding it pleasing, there is built up in us, little by little, an emotional element which gives it value, and so the love of it becomes a high, pure power which moves the self whenever any new manifestation is discovered. In like way the love of the Beautiful, and of the Good is developed.

In the beginning we have no knowledge of what is beautiful, or what is good; and so, of course, no feeling for either. We have simply the capacity for this, as for any other order of knowledge, varying in degree for different persons; and the development proceeds as a reaction to proper stimuli, only remembering that in the vital functions which have to do with purely sensuous feeling, the response is immediate, whereas, here the understanding must first construe, and then the sensibilities respond. Thus it is that nature has made us directly answerable for the degree and nature of our esthetic and moral emotions. If the understanding does not concern itself to discover what is beautiful and what is good, there will be no response of the heart upon the presentation of either the one or the other. There is no one, it is true, so low in the scale of being as to be absolutely destitute of a sense of beauty and of right; but there is a wide range in degree, chiefly due to cultivation. Especially is this true of the beautiful. In moral feeling the scale is much narrower, and probably begins lower down; but the general truth is the same. If one has no soul for beauty, it is because one has failed to open one's eyes to nature, and to study her wondrous ways; and if one is insensible to the charms of art, it is because one has neglected, through necessity or inclination, to study the works of genius. It is to be noted, however, that the intensity of esthetic feeling, with respect to what is really beautiful, is by no means strictly in the ratio of the development of the understanding. One may feel strongly, what one takes to be beautiful, though it be decidedly vulgar when brought to the test of an educated standard. We have in such case a powerful native susceptibility, but a lack of knowledge, and so a lack of refinement in the emotion. The intellect has given the object a false value, and the emotive-factor has simply accepted the purported worth. The self is then powerful, but not true. That there is an actual standard of beauty, notwithstanding the wide range of popular differences in appreciation, is shown by the fact, that all who set themselves to study the subject — all who have the opportunity and inclination to cultivate their taste in nature and art — steadily ascend towards an agreement; and although there never is entire accord in the higher plane of culture, the differences become more and more minute, with a vast range of thorough agreement with respect to the lower end of the scale.

With regard to the Good, the difference between the lowest and highest limits, is far less than in the True or the Beautiful. This is, perhaps, because of the far greater sharpness of the psychical factor which gives it character. Cognition and sensation, which respectively characterize rational and esthetic feeling, spread themselves over the whole psychical area, in a continuous and progressive way; whereas will, never absent, it is true, is in its nature decisive and instantaneous. More than this, morality, while discoverable only in action, is by no means discoverable in all actions, but is confined to such conduct as concerns the rights and well-being of our fellow-men, and our relations to the All-Father. Thus it is that the area in which differences of opinion obtain as to what is moral and what is not, is greatly narrowed; but, perhaps, on the other hand, it is just because of this comparative definiteness that its universality and depth are increased. As a matter of fact, there are no races of men, and no individuals among men, in possession of ordinary psychical activities, without a fairly definite standard of morality, and some corresponding development of ethical sensibility. Men everywhere, and in all time, have the notion expressed by the word 'ought,' in which a necessary idea of obligation is imbedded. It is the feeling lying behind it, which gives value to the particular action towards which the 'ought' points; and it is just in the right development of such feeling that man's nature reaches its supreme excellence. Furthermore, it is in this one psychical mode alone — the volitive — which gives character to moral feeling, that man in any right sense has the slightest possible power. It is only by means of this self-determining factor that the apparently independent action in sensation and cognition discovers itself; they being, as such, absolutely bound. And so it appears, that this highest possible excellence and worth to which man can attain, is exactly that which depends primarily on consciously directed self-activity;

that is to say, what he feels he ought to do is just what he is free, beyond all else, actually to do.

In the nature of the case, the self must be called upon to recognize the fact of beauty in the world, and cannot well help building up an esthetic power in the personality; and so also with rational feeling: and while it feels, in cultivating a higher and truer esthetic and rational feeling, that it is, in so far, elevated and refined, it does not feel it to be a duty, such as brings with the neglect a criminal condemnation, even though it wholly fail to strive for such elevation of nature; whereas, in moral action, a factor entirely peculiar to it must be recognized, which assumes to command absolutely, 'Thou shalt' or 'Thou shalt not' (the 'Categorical Imperative' of Kant), whenever an act involving right or wrong is in issue. This is undoubtedly the reaction of the moral feeling, giving especial value to moral action; but why should there be this altogether peculiar factor acting with an authority which we have no consciousness of having committed to it, but which we do not in the least question? The only answer which seems to meet the case fairly is, that it comes from the same source whence we are given consciousness, memory, and imagination; and as they are clearly for an end, this must be also; that end being plainly to impel men to recognize and pursue the paramount end of conscious personality.

CHAPTER XXII.

THE WILL.

Elementary effort. Emerges in conscious volition. Much that is commonly accounted free, mechanical. Liberty restricted to purposive epoch. Inhibitory functions. The office of the will in developing emotional nature. Development of volitional powers. Moral aspect of the will. Penitence.

FROM the beginning of our inquiry we have been compelled to assume a mode of the personality which makes self-movement possible. The discovery of meaning would never, of itself, be more than a dry, cold, impassive fact. It could have no value, and no end, so long as it remained pure meaning, if that were possible. To pure feeling, assuming such a thing possible, we should have to concede value, it is true; but however intense it might be, there is no power of conceiving it as doing anything as such. The discovery of meaning makes plain how or why a certain course of action would have a certain result; and feeling makes such possible result desirable; but neither of these are more than modes of the self, and we cannot conceive of a mode in itself, doing anything whatever. The ability, the power, the instrument by which the subject enjoys a capability, is still but an instrument, and cannot act of itself to make or unmake anything; the subject must put the capacity to its use. The subject which enjoys the capacity for the discovery of meaning, and of being affected by the action of stimuli, *i.e.* of knowing and feeling, is the self; but the mind does not construe, and the feeling does not feel: the self construes through the mental powers, and feels through the sensibilities. But now, though the self did understand, and did feel, the results would but be its states, and would stand apart forever, if the capacities of the self ended with these. We may conceive of the self as knowing the feeling and in some way, perhaps, feeling the knowledge; but, unless we grant a further capability, there could be no other result than that the self should simply go on knowing and feeling. Hence the absolute demand, if we are to have self-action at all, for a power to govern and control the vasco-motor system; and such power we indisputably know in consciousness, and have been all along compelled to recognize. This is the conative mode of the self. Its characteristic is energy; its office is to compel the actual to emerge from the possible; and its end, self-development. In this mode of the self we have an intuitive apprehension of causation, — of one thing compelling another. Through it is opened up to us the whole domain of the 'Becoming,' and the psychical-nexus between the actual and possible, as a simple fact, is discovered to us. All self-ordered change — every self-determined and self-enforced modification of the personality — is accomplished through its instrumentality. By cognition we are made conscious of what is; by sensation, how we are affected by the existent; and by conation, we fit ourselves to the discovered conditions.

As we have said, in this ordering of self to meet new conditions, and the consciousness of a compulsory nexus in the process of such adaptation, we have an intuition of cause. In the consciousness of effort throughout the process, we have an intuition of power or energy. The projection of these notions into all cosmical order, gives rise to these concepts as universally necessary in the mechanism of nature; and thus it must be conceded that both cause and energy, as manifested in the cosmos, have their ground in personality.

But as sensation in its basic sense is not yet entitled to be called feeling, and cognition in its elementary form is only an adumbration of specific thought, so conation, in its most general sense, is not will. All three of these fundamental self-modes are active long before the self is illuminated by the light of consciousness; that is, all along through the primitive sub-conscious period: and in their elementary forms they continue to exert their activities even in the highest stages of self-directed life. Feeling gains its title when the self enters upon the conscious use of the sensibilities; and thought claims recognition when the self uses the cognitive power to think; and so in like way, will is finally differentiated, and entitled to its name, only when it consciously enters upon the use of, and actively employs, the conative, — the energizing, causative mode of the personality.

This, rightly understood, clears the way of much popular error concerning the will. The far greater part of our actions are not volitive. The conative mode of the self has its automatic element as much as the other two modes. Indeed, as we have seen from the beginning, any automatic action carries with it all three of these as factors. If the question presents itself as to whether I shall remain where I now am, or walk into the next room, I feel that I am at liberty to will to do it or not; but from the moment I determine to do it, the movement which results will be almost wholly automatic. It is not at all that I am free to go to the other room or not; but whether I am free to will to go or not. From the instant the determination is arrived at, the movement, while due to the conative mode of the self, is purely mechanical; except in so far as any new determination may enter it. The mandate of the will no more raises my arm than the 'heave' of a boatswain raises an anchor. This self-determining activity is the Will.

All bodily movements in any wise dependent upon volition, and now become automatic, were in their inception volitive, — the will, so to speak, standing by to issue command after command, until the mechanism was taught to do its work, and the will freed for higher duties. It must not be understood, of course, that the whole work was, or even can be, volitive purely. Nothing whatever could have been accomplished under such a condition. The will is, if we may so express it, the trigger of the conative mode, obedient to the finger-touch of the self, and its proper functions begin and end with such touch; but its effect is to set free, or check, the energizing mechanism of the conative mode in certain specific directions. Without this mechanical factor, gratuitously furnished in the beginning in an inchoate form, no possible work could be accomplished.

But it must also be carefully borne in mind that the personality is not originally dependent upon the volitive activities. All that vital mechanism, unknown to and out of the reach of the conscious self, must be pre-supposed; and though silent and unthought upon, it never for a moment ceases to make possible the higher stages of personal development.

Now, among these vital functions, and long antedating anything like clearly construed purposive control, we must recognize a world of appetites, desires, impulses, and propulsions, which push us on into life, and continue to sustain us in it. In and through this whole domain we must recognize the inchoate volitive mode. It is impossible to fix upon the epoch at which purposive action begins, and so impossible to date the accession of the will to the enjoyment of its specific prerogatives; but it reaches far down into the depths of propulsive development. Until the will really enters upon its purposive functions, self-development can hardly be said to begin. All that has gone on previous to that epoch, in its purely purposive aspect, must be credited to the same power which brought into being, and still sustains the vital organism itself.

These sensuous impulses still continue in the full light of consciousness, and unless inhibited throw into movement the muscular mechanism. The self soon learns that these blind impulsions often result in distress and loss; and that it is vested, within limits, with the necessary inhibitory mechanism to modify, direct, and control these organic and propulsive movements. The function of volition is to throw into action the inhibitory mechanism; and, just so long as this restraining or detached state continues, the impulsive movement is checked or defeated. The impulse cannot disregard or override the annulling or impeding mechanism so long as the self does not, so to speak, touch the lever to throw out of gear this inhibitory machinery. The prerogative of the will is, in this regard, supreme, provided of course that the impulse is one fully amenable to volition. There are many, as repeatedly said, not so answerable; and these will go on to produce movement; but the self is provided with indirect means of controlling even those, as, for example, by flight, in the case of most solicitations. Some, as a desire for food, being vitally necessary, never yield at all.

The first work of the will, then, is to restrain, direct, and refine the original impulsions of the vital organism, — a work never fully accomplished, but continuing to the end of life. The strength of these impulsions, in the beginning, is due to the organism as inherited; but by use, abuse, or suppression through volition, they are constantly modified. An original impulse is blind, — an instinct; but after a time, its end is comprehended, and approved or disapproved, given value, and it becomes properly a desire or aversion. It is the cognitive factor which lifts it out of the list of impulses, and places it in the category of desire. It has now become a trained servitor, to be intelligently used in the higher work of the personality; or, if neglected, a refractory and heady dependent, often riotous and troublesome. This work of developing the energies of the organism is the immediate result of self-determination through the proper use of volition. The lever, by which certain sets of energies are thrown in or out of action, is the will in the hand of the self; and in the ratio of the decision and firmness in its use, the results will be excellent.

So far, we have been speaking of the action of the will in regard to sensuous impulsions. These, as we have seen under the head of Feeling, are esoteric in their origin: we now go on to consider that exoteric class belonging to the higher stage of development called Rational, Esthetic, and Moral Truths.

The process here is just the reverse of that we have been considering. In sensuous feeling, impulses blindly propel, and the sensations are themselves immediate objects of cognition; they must acquire an intellectual factor before these can be subject to volition, and then the action is directory or repressive. But material of thought coming from without, does not present itself to consciousness in its sensuous aspect, but as an intellection. There is thus primarily nothing to restrain, because there is nothing propulsive. It is not until after the meaning disclosed is transferred to feeling, and given value, that emotion discovers itself, and government becomes necessary. But passing the conative activity necessary in order to make perception clear and definite, there is very definite and positive work for the will to do in directing and holding the organism upon the subject to be studied, either in a rational, esthetic, or moral aspect; though of course the case will differ through many degrees for different organisms. In this higher stage, the instrumentality is the same; but the work of the will becomes essentially positive. It is no longer inhibitory and regulative, but impelling and upbuilding. Little by little, the meaning of things disclosed to the understanding as true, or beautiful, or good, is, by the active agency of the will, transferred to the emotive mode; and thus if the work is well done, there are built up, on the mechanical side, new and high desires, — the whole emotive nature becoming refined, enlarged, and purified.

Now the self, from the aspect of its manifoldness, is the scene of constant strife, — impulses conflict, desires conflict, — and one is torn by contending factions. Order can only be maintained by the dominating agency of the will; and, under ordinary circumstances, peace obtains; but sometimes, under unexpected and trying circumstances, the will is vacillating and unsteady, and the whole self in a state of confusion and uncertainty. Few persons but at times suffer such a stampede; but some are so habitually changeable and unsteady that they have no confidence in themselves, and cannot be depended on by others. This is because they are constantly throwing the conative organism in and out of gear; or, in consequence of a confusion of the understanding, are in perplexity as to what connection should be established and maintained. Thus the understanding is ever presenting different courses as possible and promising, and feeling is clamoring for the gratification which lies in certain directions, while prudence gives warning of the danger or loss which impends: the self must decide — must choose between the two or more possible courses of action: and the will is the instrument of the choice.

As a simple fact of consciousness, the self experiences no possible compulsion upon it to choose any one out of the many courses of action presented; and it is in this fact, and only at this point, that freedom can be predicated. Determinism is also a fact known only primarily in consciousness, and absolutely necessary to the notion of freedom; but determinism is felt to be in the mechanism alone, and after the will has acted. If the organism were not compelled to body forth the fiats of the will, it would be inconceivable that any effects should follow a cause; and so, that any such phenomenon as volition could ever be. Thus we arrive once more at what we have had abundant reason to see before, the personal and conscious origin of cause, and the dependent uniformity and inflexibility of mechanism.

The self must ever be at its post, so to speak, to give direction to the mechanism, much as a driver must make his touch on the reins felt, though hardly conscious at most times of the effort. In the far larger round of routine life, desires run so evenly with the purposive trend of the self, that the mechanism seems to act of itself,—and, indeed, it is the objective self acting; for those settled desires with respect to business, society, study, and whatever else that gives tone and intensity to our daily walk, have all been built up by innumerable conscious efforts of the will in the past; and now, by their stored-up energy, are carrying the personality forward, through these self-developed powers. But there is always considerably more of the original and self-directing factor, even in our routine life, than appears; much as with the driver, though he appear to exercise so little control over his horses, if he were to let go the reins for a moment, the results would ordinarily show a marked difference in consequence.

The functions of the will are scarcely less important in the work of moulding and cultivating the intellect. It is the province of the understanding to disclose the order and bearing of things and events, and this it does according to its excellence at any particular moment; but it is a psychical mechanism, and as such is not only susceptible of large variation in accuracy and power, at every stage, but needs to be developed from the beginning. The office of the will in fixing and directing attention from the beginning of the thought-process is of the highest moment; and without it there could never be anything more than a chaotic thought-mechanism. Assuming a fair development of intellectual power to have been already accomplished, so long as our thinking proceeds along accustomed lines, and on accustomed subjects, the will has little more to do than it has in its ordinary government of the vasco-motor system; but whenever an intellectual difficulty arises, or where the subject does not call into activity the stored-up energy of ready-formed desires, the self is compelled to hold the mind hard upon the point in hand, and stoutly refuse to yield to the weariness or indifference which is sure to ensue. If the self keeps the intellect steadily applied, as a rule, difficulties begin to yield, and indifference gives way. There are undoubtedly great differences in the native powers of the intellect; but the high things accomplished in the thought-world are far more due to the persistence with which the self uses its intellectual capabilities than to their original vigor. Sir Isaac Newton used to declare that he succeeded in his work, not by any extraordinary sagacity, but solely by patient and persistent thought : " He kept the subject of consideration constantly before him, and waited till the first dawning opened gradually into a full and clear light, never quitting, if possible, the mental process until the object of it were wholly gained."

We may be permitted to doubt, in his case, that this was all; but the result of this process, if not carried to such an extreme as to work injury in other respects, is not only that the subject of thought becomes luminous, but an intellectual momentum is stored up in the thought-mechanism itself, which enables it to do easily what is, at first, wearisome and difficult. This is the true work of the student, that is to say, of one whose object is the development of his intellectual capabilities for subsequent work in special lines of investigation, or for the general work of life. The materials of thought — the actual information and learning gathered by the way — are necessary to the process; but, after all, cannot be compared for value with the acquirement of mobility, accuracy, and intensity in the thought-mechanism itself. If, on the other hand, the self fail to use this marvellous will-power in a true process of intellectual evolution, difficulties gather rapidly from the very neglect or misuse, which finally defeat better efforts made too late. Rust, obstructive accretions, and undertows of habit, clog and impede any serious effort of the self to direct the current of thought from old channels, and to hold it long enough in the new direction to wear its way along higher lines.

But not only has the self this power of refining and strengthening the intellectual mechanism through the right use of the will, but it has the power also of developing and strengthening the will itself. The process is this: the understanding makes one conscious that one is too soft and yielding to the pressure of desires, to solicitations from companions, and to difficulties in the way of study; the feeling gives value to the disclosures of the intellect, and one resolves (an act of will) to be more firm in future. A new occasion presenting itself, one is but partially successful, but renews the purpose; a little better success the next time, and so on to decided success. If one knows oneself to be vacillating and unsteady in affairs, one can practise decision in small matters, and firmness in adhering to determinations, though at some cost or annoyance, and if honestly conducted, the will can be trained as certainly as the intellect and feeling. Self-discipline, in all the modes of the self, is of the utmost importance; and it seems a pity that young people do not understand better the psychological principles underlying it.

A man's acquisitions in this world fall into two great and vastly different classes, — those which come to be his subjectively and organically, and those which come to be his objectively and artificially. Thus the strength of muscle, the sagacity of mind, the decision and honesty of character, all belong to the personality, as organically real and living factors, — they *are* the self manifested. All acquisitions of this class can be gained only by the co-operation of the self, through the will.

The acquisitions belonging to the second class are those which come to one from without, and have no vital and organic ligament connecting them with the personality. They are such as wealth, station, honors, friends, and whatever else one may claim, in a proprietary sense, of the eternal world. These may be bestowed, purchased, seized; but the nexus between them and the self is never more than a fiction, and they may be totally lost at any moment. The difference between the two classes is like that between fruit actually growing upon a tree, and the tied-on fruit one sees at Christmastide for the children. Now, this real and true fruit of mind and heart and will can be had in no way but by and through self-effort; and the supreme, and dominating factor, through it all, is the will. No money, nor influence, nor favor ever yet gave one the power to read or write, to play upon an instrument, or solve a problem in mathematics. Money and favor can, indeed, supply the opportunities; but no muscle will ever do its office, no brain-cell will ever exercise its cunning, except by the will of the one only person who shall be able to actually enjoy the gain. And so, in all those higher reaches of self-development. The only real value of all that second class, for the poor, artificial possession of which there is so much zeal and strife in the world, lies in its bearing upon, and use towards the acquisition of the true possessions which come to us through the self-developing realities of the subjective class.

The most important relation of the will to human destiny still remains to be mentioned. It is its moral aspect. All philosophers are agreed that there is an important difference between conduct which we call prudent, agreeable, or sagacious, and conduct which is morally good. The difference practically between an action which involves the question of right and wrong, and one which is simply prudent or foolish, is understood by everybody, and needs little explanation from the every-day point of view. Theoretically it is an ethical question, and I shall defer this aspect for consideration at a later stage of our inquiry. Here we confine our treatment of it to a brief look at its psychological bearing.

Obviously the will is an essential factor of all questions of conduct, either prudent or moral. But in every act, we may distinguish three stages; the conscious promptings of the emotional nature under the disclosures of the intellect, the purposive self-determination, and the results which follow this purposive epoch. Of these, there is no liberty in either the first or the last. The emotional nature must make itself known in consciousness for just what it is under the existing state of the case; and after the self has culminated in a volitive fiat, the vasco-motor mechanism must execute the mandate with the same absolute necessity as one cog-wheel compels the movement of another. It is inconceivable, therefore, that the self should hold itself accountable for the solicitations which assail it under an existing environment, though conscious of the fact that it has had much to do with building up, in the past, what are now facts of its own nature. So also, after the purposive epoch, the results, falling under the law of determinism in nature, no sense of accountability in them, as such, can be recognized by the self. But consciousness has quite another voice with respect to the purposive epoch. However strongly the self may know itself to be impelled to a particular volition, consciousness is clear that the decision rests with the self alone; and a corresponding sense of accountability is recognized. It is therefore the purpose alone which can give self accountability to an action. If the purpose be good, the act is good; if consciousness recognizes the purpose to be bad, the act is bad. We do not now inquire what the ground for this recognition of moral quality in purpose is: we here confine ourselves to the bare fact that there is such recognition. As to whether the purpose with which the will acts is or is not good, the self is the supreme and only judge; and it, and it alone, can know whether, in the light of the understanding at the moment, any particular purpose has this moral quality or not.

In many, indeed most, actions this moral quality does not appear at all, and the reason of this fact will appear when we come to this inquiry. Such actions fall under the general head of prudence, taking the word in its largest sense. Such actions may be either those which are intended to promote general well-being, by the accumulation of power in the shape of wealth, or education, for future use; or those which give immediate pleasure, and have no further end. The success or failure of one's purpose to accomplish the object, with all dependent consequences, will determine whether the particular act is wise or not. Thus it is that a prudent, or an imprudent, action depends upon what happens after the purposive epoch; and this fact broadly marks a distinction from all moral action: — prudence being justified by the mechanical results after the act of volition; morality, by the quality of the purpose itself. Bad results make known the unwisdom of the acts, and one may be sorry that one did not act differently: a good act may also result unfortunately, but one cannot feel guilty, however much he may deplore the event, or where, under a mistake, he had not understood the case better. Once assured of the right purpose of an act, in any state of case, one cannot be sorry he had not acted differently, unless he regret that he was true to his own consciousness of right. There can never be penitence, unless there is a clear conviction that the act had a guilty purpose, however much one may bewail the result. Here we leave the question for the present.

CHAPTER XXIII.

UNITY OF PERSONALITY.

Difficulties of question. Unity and manifoldness. Unity a primordial condition. Inferior organisms. Protozoa. Not two worlds, one spiritual and the other physical. Man a manifestation of one person in two hypostases. The psychical and mechanical inseparable. Gross and sublimated matter. Visible and invisible universe. The 'Unseen Universe' quoted. The mechanical mode has its title to reality only through Personality. The cicada.

ONE has no difficulty in understanding the Unity of Personality, so long as one is content to know it for and in oneself; nor is there any difficulty in recognizing it in another. But in attempting to say, explicitly, just what it is that one means by this unity, a serious trouble starts up, and continues to confront us. This will not be a surprise, however, if we keep in mind what has been so constantly insisted upon in these pages, namely, that the ultimate is known, not by the construing power, but by the power that underlies all relations, — the power through which we derive all rational intuitions, and which is rather pure feeling than articulate thought. The understanding, concerning itself with the differentia of concepts, pronouncing upon the likeness and difference among the divers marks of an object of thought, has no difficulty in dealing with personality so long as it is permitted to seize upon and emphasize its modes, states, and conditions. These all, as elements of the empirical ego, take their place in the self's world, — the sole and only world that is, or can be, known to any one of us: — but when all marks are denied it, the construing power — the power which demands subject and predicate, and pronounces upon their congruence or incongruence — has no field in which to exercise its functions; so that, in the sphere of the understanding, this aspect of 'knowing' is impossible. The self acting, the self suffering, the self in exhibition of any phenomenon, presents material for cognitive scrutiny, and the understanding is at home in the execution of its office. But it is obvious that the necessary conditions of all such activities is plurality; and, therefore, whatever the understanding can discover of the self must carry with it this underlying condition of manifoldness. For this reason, it is hopeless, and the effort involves contradiction, to look for the unity of the self in any deliverance of the understanding.

And yet every such deliverance implies that very unity which it is incompetent to construe. In any judgment touching two concepts, A and B, the full statement undoubtedly is, I see, — I declare, — I know A to be like or unlike B. In any sense-perception, the discovery of order and arrangement, of beauty and power, the consciousness rendered explicit is, I see, I discover, I feel. The self, thus, is always presupposed as the centre of the in-and-out-go of whatever affects it, or is affected by it. This is so obvious, in view of what has been said already, that it needs no further amplification here. We are simply back once more to the domain of the Pure Reason, — the source and ground of all knowing. To be explicit, the point is this: — the understanding tells us plainly that we cannot *understand* the unity of the self, nor the unity of anything whatever, apart from its plurality; but 'to understand,' it must be borne in mind, is to be taken in its technical sense; and by no means as synonymous with 'to know.' We do know, and must know, the self in its unity as a primordial condition of all other knowing of whatever kind soever.

This brings the fundamental truth, and key-note of this work, once more clearly before us: — the co-existence of the 'one and the many.' The title of the book is but another way of phrasing this mystery of the ages. Personality as the subject knowing, feeling, acting, is essentially One; the modes of the personality as known in cognition, feeling, and volition, are essentially manifold. The knowing instrumentality, the feeling instrumentality, the acting instrumentality, are, in their severalty, mechanisms; and each in itself is a plurality. The self, as subject, has a proprietary right to its modes, and they must look to it for their reality; and so the entire mechanism, in its physical and psychical aspects, belongs to the personality in its unity. We say 'my thought,' 'my feeling,' 'my will'; and we never for a moment mistake any one of them, or all of them together, for the self. They are the manifestations of the self, as much to the self as to others. Thus, while they are not non-self, they can have no possible title to be at all, except in and of the self; and so cannot be thought of as separate, or separable from the self. Neither, on the other hand, can the self be thought of as separate, and apart from its modes, — those now known in consciousness, or others which may replace them. This, as the metaphysical problem of substance and quality, we shall have to speak of further on.

All this applies to the whole bodily organism. The body is not the self, but it is its physical manifestation, absolutely necessary to it; and so, inseparable from it. But when we speak of the body, we do not mean an individuality incapable of numerical distinction and separation, but we mean a plurality made up of divers parts, all undergoing constant change. The body is not the same in its physical composition at any two consecutive moments. The body, necessary to any thought of the self, is thus not an identity, — not this now-existing body as it is in every particular, — but *a* body is necessary, in which the modes of the self shall continuously find their mechanical basis. Thus the body may lose a limb, any amount of tissue, even large masses of the brain; but consciousness does not tell us that the self suffers mutilation, or is less itself than before.

This is because the body is, in its very nature, a plurality, made up of parts and susceptible of objective increase and diminution.

We have in this fact a notable distinction between body, as such, and the personality in its bodily manifestations. When a mere body, or thing, is increased, diminished, or divided, our thought is that *it* is increased, diminished, or divided; but we never think, and cannot think, that the personality suffers accretions, subtractions, or dismemberment by any possible process which may be performed upon it physically or psychically. We do indeed think of it as changeable and changing; but never in such wise as to discredit, or in any possible way touch its unity.

Undoubtedly 'thing' may claim a sort of unity. It is 'one' thing, and of this it cannot be defeated; but it is a unity that at once becomes two upon division; whereas, we cannot form the least conception of what could be meant by the half, or any fractional part of self. This is true also of anything having sub-human personality, as a brute; or of whatever has an animate existence at all, as a vegetable. We are compelled to recognize a vital oneness which is not susceptible of fraction.

Science, it is true, acquaints us with innumerable examples of organisms in which it cannot be denied that there is a oneness of vital structure, and yet, upon a section being made, the dismembered part readily acquires an independent organic existence possessing all the characteristics of the parent organism. "The severed parts of the mutilated polype become wholes by growing into perfect animal forms, in each of which is fully evolved the sum of psychic capacities that belong to the original uninjured creature." Morphologists are agreed that the protozoa exhibit endless examples of composite life, — states of existence in which a paramount individuality is exhibited, and yet which are composed of innumerable sub-individualities; even rising through several orders of integration, and composing individualities of greater and greater complexity; as in the sponges and the hydra. In many animal and vegetable species, propagation is accomplished by the spontaneous severance of portions of the parent body; such portions developing into the perfect organism of the species. Substantially the same phenomenon shows itself even in human generation.

Obviously it is impossible for either naturalist or psychologist ever to know more of the nature of this unity in complexity than appears in its extrinsic phenomena; and it is safe to assume that there will always remain a wide domain of conjecture in any attempt to explain the organic nexus. It appears plain, however, that when an artificial severance is made in such an individualized community, it is not a matter of complete indifference where the section shall take place (so long, at least, as we can, for minuteness, distinguish the sub-units), but that, for the new community life, or an isolated individual development, the severed part must carry with it a complete germinal unit of the being to be developed. If there be, indeed, a true unity of the first, or any other order in such personæ, the several sub-personalities but constitute the mechanism of the paramount individuality, and a section would not cut in two the personality of (say) the polype, but simply cut the corporeal bond which holds one part of its vitalized body to another; giving to the severed portion the opportunity to enter upon an independent existence, and develop a more perfect organism than it could before enjoy. In point of fact, except that we do not see the development of independent creatures, the same thing takes place upon the separation of a portion of living tissue from any one of the higher animals, man included. Physiology does not permit us to doubt that our bodies are aggregations of an infinite number of protoplasmic units; so that, in reality, the morphological facts of the protozoic and polyzoic worlds present no greater difficulties on this point than are found in the human organism. We but find ourselves struggling with dissonances which, without yielding up their antagonism, are welded into a broader and deeper harmony; or with individual motions, as of the earth, free and independent in their several spheres, composed of all velocities in all directions, swallowed up and lost in the larger sweep of her paramount orbital individuality.

But, whatsoever the external world reveals to us, we must ever remind ourselves that it can never acquire any better warrant for its reality than is given in consciousness; and to this at last we must fall back — as from it, at first, we were compelled to set out — in any discussion of the unity of personality. But here some caution is needed. It is not because we appear to ourselves in consciousness to be one, that we get our guarantee that we are actually one; for, manifestly, many things appear, and persistently appear, to us which we know to be semblances; but, as Lotze says: "Our belief in our personal unity rests not on our appearing to ourselves such a unity, but on our being able to appear to ourselves *at all*. Did we appear to ourselves something quite different, nay, did we seem to ourselves to be an unconnected plurality, we would from this very fact, from the bare possibility of appearing anything to ourselves, deduce the necessary unit of our being, this time in open contradiction with what self-observation sets before us as our own image. What a being appears to itself to be is not the important point; if it can appear anyhow to itself, or other thing to it, it must be capable of unifying manifold phenomena in an absolute individuality of its nature."

Insisting, then, upon the unity of the human personality discovered to us as a primordial intuition, and supported at all points by the necessities of thought, we are compelled to recognize the self as that in which and for which all its modes are entitled to reality, in any ultimate sense. It is thus itself the primordial reality, and is not to be thought of as an abstraction, a somewhat vague and shadowy, which may be set loose from its modes, and still exist destitute of quality or relation, out of space- and time-forms, as, indeed, pure and simple existence, which is pure and simple nothing. Some such vague notion has been, and still is, the popular teaching on this point, — the legitimate and necessary outcome of that dualistic theory, reaching, in modern times, back to Descartes, in which mind and matter are held to have no possible touch, or point in common, — to belong indeed to two substantially different worlds.

In our view, there are not two worlds, one psychical and the other physical, but one world of which these are the two modes; both real, but with a common ground deeper than themselves, the Ultimate and Infinite 'One,' — the Self-Limited, Self-Existent Personality.

This carries our thought back to the analogy presented in the lower forms of life, in which we see paramount individualities gathering up into unity the manifold sub-personalities of their organisms.

Man, the microcosm, is one self, manifested and realized in his two hypostases, one a physical mechanism, and the other a psychical instrumentality. He is thus not a mere physical being, nor a mere psychical being; but a physio-psychical being, — a *person*, gathering up into himself the innumerable limited self-activities of his bodily organism, himself a limited self and a self-limiting reality.

According to our view, it is a mistake to hold that the soul, or mind, or spirit (whichever one of these may be preferred as the name of the psychical mode of personality) can possess an independent and separate existence, any more than the physical mechanism could possess a unity apart from the self. The empirical self implies, and absolutely demands, the psychical factor; and this would be lost by the destruction of the physical basis; and so, also, any activity or energy of the self demands that it shall have modes by which psychical differentiations shall take place, and these modes cannot, therefore, be detached from the self, nor the self from them.

Unless, then, we mean by 'soul,' or any other like word, to embrace the entire personality in its essential physical and psychical oneness, it is misleading and inadmissible; such words can be used in any other sense to designate, not the self, but some mere mode thereof. We say 'my soul,' 'my mind,' 'my spirit,' etc. In this sense, the use is correct and clear enough; but any meaning which implies a possible diremption of the personality is erroneous and unjustifiable.

But, it will be asked, if both the substantial and mental modes are absolutely necessary to the integrity of the self, what becomes of personality at death? The question is pertinent and demands an answer, though it might be better in some respects to defer it until after we have seen what light advanced science is able to throw on the subject of the essential nature of matter. It is necessary to so far anticipate the results of the discussion of that subject as to say that our ordinary conception of gross matter, as a somewhat hard, impenetrable, or visible, is the result of an invincible prejudice. We shall see that there are many forms of substance which offer no sensible resistance, are not visible, nor in any wise discoverable to our present senses; some, familiarly known to us, as, for example, water, which are solid, liquid, or gaseous under different conditions; while with that inter-stellar ether which pervades all space there is no semblance of sensible qualities left, not to speak of the possible 'perfect fluid' of the advanced physicists. There is then nothing in physical science against, and much that absolutely demands, the hypothesis that all matter in its ultimate analysis is super-sensible and invisible; so that we are not only permitted, but required, to hold that the gross matter of our bodies is but one out of any number of possible forms which it may assume under different circumstances.

The authors of the "Unseen Universe," themselves leaders of scientific thought, say they are led to conclude that "the visible system is not the whole universe, but only, it may be, a very small part of it; and that there must be an invisible order of things which will remain and possess energy when the present system has passed away. Furthermore, we have seen that an argument derived from the beginning, rather than the end of things, assures us that the invisible universe existed before the visible one. From this we conclude that the invisible universe exists now, and this conclusion will be strengthened when we come to discuss the nature of the invisible universe, and to see that it cannot possibly have been changed into the present, but must exist independently now. It is, moreover, very closely connected with the present system, inasmuch as this may be looked upon as having come into being through its means. Thus we are led to believe that there exists now an invisible order of things, intimately connected with the present, and capable of acting energetically upon it, — for, in truth, the energy of the present system is to be looked upon as originally derived from the invisible."

Upon the subject immediately under consideration they go on to say: "Let us begin by supposing that we possess a frame, or the rudiments of a frame, connecting us with the invisible universe, which we may call the spiritual body.

"Now each thought that we think is accompanied by certain molecular motions and displacements of the brain, and part of these, let us allow, are in some way stored up in that organ, so as to produce what may be termed our material or physical memory. Other parts of these motions are, however, communicated to the spiritual or invisible body, and are there stored up, forming a memory which may be made use of when that body is free to exercise its functions.

"Again, one of the arguments which proves the evidence of the invisible universe demands that it shall be full of energy when the present universe is defunct. We can therefore very well imagine that after death, when the spiritual body is free to exercise its functions, it may be replete with energy, and have eminently the power of action in the present, retaining also a hold upon the past, inasmuch as the memory of past events has been stored up in it, and thus preserving the two essential requisites of a continuous intelligent existence."

This is sufficient to assure us, negatively, that there are no scientific grounds upon which to conclude that the dissolution of the gross and palpable body deprives the personality of its essential physical mechanism; but, on the contrary, that there is better reason to hold, positively, that the truer impalpable super-sensible 'spiritual body' still continues, and retains all the energies and potentialities possessed by the sensible body.

But as we have seen that the mechanical mode of the self has its ground and only title to reality in the personality, — and in this we shall be still further certified as we proceed, — we are unable, in any way we can make articulate to ourselves, to conceive of its discontinuity or dissolution. That we cannot positively construe to ourselves the state or condition of the personality after death, must be admitted, but it must be remembered that such state or condition would be an actuality, and that we are in no worse case with regard to this than we are with respect to any other possible unexperienced fact. We cannot in advance of experience imagine even a new flavor, a new perfume, a new state or phenomenon of any sort whatever, even of this present world, though we are quite sure many such facts of experience will, in the course of time, be revealed to us. All we can do by the powers of the imagination, as we saw in discussing that psychical energy, is to combine and arrange, under the laws of thought, the materials already in our possession, by experience; but never by any possibility can we introduce any absolutely new element into a fabric of fancy. Now, of a super-sensible world we have no possible experience, and, with our present powers fitted to the needs of a visible world, we never can have any; and thus shall never, in this life, be able to gather the least material through which to construe such a post-mortal existence. It follows, therefore, that we are, by the necessities of the case, utterly incompetent to form any conception of the state or condition of a super-sensible existence; and so our failure in this regard in no wise makes against the probability of such a future state.

The case then stands thus: before we could conclude that the self—the, to us, one and only absolutely indisputable certainty—suffers or could suffer annihilation, we need to have the most overwhelming evidence; whereas such evidence is in no particular forthcoming; but, on the contrary, we have much proof against the hypothesis and in support of the conviction among all men and in all ages, that the human personality does not terminate at death.

All nature is full of exemplifications of bodily transformations. One example out of the multitude will be sufficient for our purpose, and I fix upon a species of the cicadæ—the 'seventeen-year locust.' Passing over the twenty-five or thirty bodily transformations which it undergoes during its long larval existence underground, it at last bores its way to the surface, and appears above ground, completely encased in a hard, horny body, with legs, antennæ, and all the paraphernalia of a creeping bug. It makes for the nearest tree and begins to climb, but does not get far before an astonishing change takes place. The shell splits down the middle of the thorax, the mask (larva) is left behind, the pupa passes into the perfect cicada, emerging an altogether different creature in function and appearance, now with iridescent wings, an astonishing instrument for the production of sound, and everything ready prepared to enter in a moment upon a sphere of existence wholly new and unlike any of the many through which it had previously passed. The shell or mask, with all its legs and digging apparatus perfect, looking still very like the creature of which it is now but the exuvium, still clings to the tree, while the real creature, in its new mechanism and new splendor, spreads its wings in the aërial world, and sets up its ear-splitting song.

Now in this, and in innumerable like larval changes, the transformation is, it is true, from one form of gross matter to another; but it is clearly unscientific to urge an objection which turns upon a mere degree of refinement. Substitute for the visible substance of the emerging pupa one of the many impalpable and invisible forms of matter, and we should have some far-away, but doubtless quite analogous, transformation, to the emergence of the spiritual body from the 'mask' of the sensible, common form of matter which encloses our more perfect bodies, fitting them to our present uses, while they are preparing for a higher, hyper-sensible existence.

CHAPTER XXIV.

WHAT IS 'THING'? CONSTRUCTION OF MATTER.

Illusions of nature. What underlies phenomena? Pure Being. 'Thing' that which affects and is affected. The position of Bishop Berkeley. Quoted. Analytical physics and construction of matter. Boscovich's theory. Molecular Mechanics. Clerk Maxwell. Professor Tait. Sir William Thompson's vortical atom. Difficulties. Le Sage's theory. Ether. The physicists driven into metaphysic. Atoms 'manufactured articles.'

NATURE gives us constant warnings not to accept appearances for more than they are worth. Whatever appears is, in so far, true; but we are expected to learn that there is always a deeper truth for the sake of which the appearance is, and for which, indeed, it must be. The appearance, though not real in its own right, is not a delusion, but becomes one when we stop upon it, and insist on taking it for an end in itself. Thus illusions, by our own fault, constantly pass into delusions.

There is no just ground to complain of being deceived by the appearances in the natural world, though there is nothing that would not serve as an example of how we accept the seeming for the real. We see the vault of heaven above us, but we know that it is an illusion. The sun and the moon, with the myriad worlds at every possible difference of distance, and moving at enormous velocities in every direction, are all apparently brought together on this spherical surface, and worlds and systems present themselves to us as fixed and glittering points. If there is one thing clear to us, it is that the earth under our feet is at rest and immovable, and yet its motions, in rapidity and variety, rival the particles of dust in the fiercest whirlwind. We see the celestial bodies cross the heavens from east to west, but we know that the movement is not in them, but in us. The rainbow hung in the clouds is not there, but on the retina of the eye.

It is quite the same in terrestrial things. The hues of the lily, its perfume, its apparent magnitude, are not in it, but us. The hoot of the owl, the warmth of the fire, the flavor of the peach, are not realities of the outer world, but are only states and conditions of our own sensibilities. What do we see of the vibrations which are the physical basis of sound, heat, electricity, light, and perhaps all material phenomena?

Then do our senses deceive us? No, certainly not. They do just what by their office and nature they are appointed to do. As Lotze says, "Color and sound are no worse because they are *our* sensations."

But the question presses itself upon us, What is it that appears? What is 'thing'? We do not ask now, how we know 'thing,' but, What is it?

In the first place, it cannot be any mere quality, nor the sum of any number of qualities. Hardness cannot separate itself from the 'thing' which is hard. The weight of a body changes as it is carried from the equator to the pole; and if it could reach the centre of the earth, its terrestrial gravity would be zero. We can form no possible notion of resistance, if there be no somewhat to resist. And so, all that list of the qualities of matter, called primary, because they depend not upon us, but upon the ultimate structure of body, presuppose a substratum — substans — or substance — something standing under and supporting these qualities. It is of this underlying something that we inquire.

Let us then make a search for it. Take any concrete object and strip it of its properties one by one; take away its color, its weight, its hardness, its figure, size, and structure, and whatever else is discoverable to the senses. Since its dimensions are gone, it must have lost extension, and dwindled to a mathematical point. It would still have place or whereness, but position is something it *has*, not what it *is*, and as a quality must go. It may still be thought of as having a sort of ghost of its past, as having once been; that must go, too, and it will then be out of time and out of space, with nothing predicable of it. It will be just what we mean by 'nothing'; and yet we are still speaking of 'it'; and in our stripping process, up to the last moment, when the last token of its qualitative existence was taken away, we had to assume the 'it' as still existing; and unless we admit, as we cannot, that quality can be substance, since we have taken nothing but qualities away, its being — its 'itness' — must still remain in its pure form; and we are compelled to say with Hegel, " Pure Being is Pure Nothing."

It is not our province to build up the universe; we have quite enough to do, to understand something of its nature and laws, now that it is existent; but if we were to attempt it, we should like some better footing to start with than the perfect emptiness — (no; emptiness is a quality or condition, and Pure Being cannot have even that) — the perfect nothingness (if that does any better) of Pure Being; but even the qualifying word 'pure' has no right to flourish before the word 'Being'; and, to have even 'Being,' is to have the attribute or quality of being, so that neither this nor any name or sign can be left it. It must not even be thought upon, — nay, not even banished from thought!

The first, and every step towards recovering reality, as we know it in the universe, from the utter nothingness of such an ultimate subject, with motor and movement engulfed of nothing, would present the absurdity of starting on a quest for what has no name, nor place, nor time, nor other mark by which it could be known; and if this difficulty could be got over and it could be found, it would have nothing to which any of the qualities it has given up could be attached. And yet this is not far from what is attempted by the Absolutists.

We do not sympathize with them in their efforts, and are quite content to expend our energies in attempting to understand something of the actual, beginning our efforts far this side of *mere* Being. Lotze says: " Reality means for us the ' Being ' of a somewhat that is capable of being affected and producing effects. Everything with which this definition comports, is accordingly called a ' reality,' — that is to say, has this title. But there cannot be a ' reality *per se* ' — which were nothing — as the bearer of this title. What is supposed to be *real* must merit this designation by being susceptible, through its own definite and significant nature, of having reality in the meaning alleged."

To come back, then, to 'thing,' we can only say that it is what by the nature of its actual existence, under the intuitions of time and space, produces effects and is affected. These effects in us we know, and so know 'thing,' or body, or substance directly. Whether substance is material or spiritual we need not ask, because the question really means nothing. Since, confessedly, nobody knows or can conceive of what matter is, we travel wholly out of the record when we undertake to pronounce upon whether it is like this, or like that. Matter is just what it is, and just as it is, it appears to us — now in one phase, now in another, all equally true, and equally real. The very nature and purpose of 'thing ' is to affect us and other things; and when we try to force another meaning, which seeks to see beneath the nature of 'thing,' and find another nature upon which its world-nature is founded, we but commit ourselves to an infinite regressus.

This is, substantially, the position of Bishop Berkeley, whom so many people, even to this day, fail to understand. " I am of a vulgar cast," he says, " simple enough to believe my senses, and leave things as I find them. To be plain, it is my opinion that the real things are those very things I see and feel and perceive by my senses." Again, "For my part, I can as well doubt of my being as of the being of those things which I actually perceive by sense; it being a manifest contradiction that any sensible object should be immediately perceived by sight or touch, and at the same time have no existence in nature, since the very *existence* of an unthinking being consists in *being perceived.*"

This he is constantly declaring, and yet, the world, in the main, will have it — great and good men insisted upon it in his day — that he did not believe in the reality of things. One is at liberty to reject his theory, but not to misrepresent what he taught. "I do not argue," he says, "against the existence of any one thing that we can apprehend either by sense or reflection. That the things I see with my eyes and touch with my hands do exist, really exist, I make not the least question. The only thing whose existence we deny is that which *philosophers* call matter or corporeal substance. And in doing this, there is no damage done to the rest of mankind, who, I dare say, will never miss it." His denial is only of that 'nothing' which we found to remain in thought after everything had been taken away from our object a moment ago. Berkeley's position is admirably put by Lewes in his "Biographical History of Philosophy." "If by matter you understand *that* which is seen, felt, tasted, and touched, then I say matter exists; I am as firm a believer in its existence as any one can be, and *herein I agree with the vulgar.* If, on the contrary, you understand by matter that occult *substratum* which is *not* seen, *not* felt, *not* tasted, and *not* touched, — that of which the senses do not, cannot, inform you, — then I say I believe not in the existence of matter, and *herein I differ from the philosophers* and *agree with the vulgar.*"

This presents the issue sharply. Nobody denies the existence of matter in the phenomenal sense, *i.e.* as a somewhat which offers resistance, and is known to the self. But the question is as to whether, if we could go on dividing and dividing, and could carry the process far enough, we should finally come to bits of matter incapable of further division or change; that is, whether there is an eternally subsisting, dead, unknown, and unknowable stuff out of which the external universe is constructed, or not. It was this which was questioned by Berkeley, and is questioned by the advanced thinkers, metaphysicians, and physicists in our day as well.

It will be well, certainly interesting, to consider briefly what is thought of the nature of matter from the side of analytical physics.

Dr. Thomas Young, than whom a more profound mathematician and physicist scarce ever lived, in discussing the theory of light a hundred years ago, came to the conclusion that the diameter of a particle in a substance of the density of water " must be less than the hundred and forty thousandth part of its distance from its neighboring particle, and thus the whole space occupied by the substance must be as little filled as the whole of England would be by a hundred men, placed at the distances of about thirty miles from each other."

Upon this hypothesis, a particle of such a substance would be relatively about twelve times further from its nearest neighbor than the earth is from the sun; or, to put it a little differently, if the particle were magnified to the size of the earth, and the distance in the same ratio, the particle would be more than a billion of miles removed from its nearest neighbors.

Since Dr. Young's day the mathematicians have done an immense work in molecular physics, so that it is now a well-established title in mechanics. We can but glance at some of the salient points touching the construction of matter.

Perhaps the most famous of all the hypotheses on the subject of the composition of matter is that put forth, now more than a hundred years ago, by Father Boscovich, a Jesuit priest.

He attempts to get rid of a difficulty which meets those who contend for the continuity of matter, and those who hold that it is discrete, or composed of aggregations of elements of such character as to be in the last analysis essential units. Those who, like Descartes, hold that extension is the essence of matter, are of the first class; and in their view there can be no separation of one bit of matter, either large or small, from another, but there is a material *plenum*, or absolute fulness, in the universe. This may be better understood by applying the notion to a substance of a homogeneous character, such as water, in which all drops flow into each other so that there is no discontinuity. The whole may be divided into drops; the drops when thus apparently separated have matter, as air, which connects them, and if these should be separated again, there is still a subtler form of matter, as ether, which maintains the continuity. Thus we may and must have the possibility of infinite divisibility, but never discontinuity.

The other school of thought hold that there are ultimate atoms, very small indeed, but still of some magnitude, and of such character as to defy all possible effort at diremption; and that these distinct and separate entities, lying far below our powers of vision, exist according to certain laws or orders, and compose what we know as substance.

Boscovich abandons the notion of any ultimate, hard, self-subsisting bodies or atoms in this sense, and holds that there are only mathematical points existing, in the ultimate analysis, which however he endows with mass or inertia, and makes capable of retaining their identity, with power of movement, and of action and reaction on other like endowed points. They are centres of 'force,' attractive and repulsive, alternating according to the distances between them; ending on the one hand in the law of gravitation, when the distance is greater than about a thousandth of an inch, and on the other, in insuperable repulsion, when one point-atom is infinitely close to another. Two of these atoms can never occupy the same place, and can never be in actual contact, since such contact would require an infinite force. Now, since there are alternations of the force of attraction and repulsion, there must be at certain distances positions of neutrality; that is, where attraction changes into repulsion, or the reverse. These are positions of equilibrium or permanence. Now, a certain number of these force-atoms, lying in every possible direction from each other, occupying positions of neutrality, must be in a state of equilibrium, and constitute a system. They will defend their ground against all other systems by the law of composition and distance, so that we thus have sensible bodies, when there are enough of such communities to produce a reaction on our sensibilities.

If the distances between these centres are changed by being forced into new positions of neutrality, either closer together or further apart, the nature of the substance is changed; so that the differences between all material substances are accounted for by the more or less remoteness of the assumed centres in its constitution, and are not at all due to any essential differences in the original stuff.

This theory at first blush seems wild enough. But as modified by the latest results of molecular science, it stands its ground; nor do the physicists, in a general way, and with an exception to be noted, seem to see or desire any escape. Accepting Professor Clerk Maxwell as authority, we shall follow him chiefly in stating the assumptions and conclusions of molecular mechanics.

First, bodies are made up of parts, each part being capable of motion, and these parts act and react on each other in consonance with the principle of the conservation of energy. All these parts are in motion, rest being only a particular case of motion. "The phenomena of the diffusion of gases and liquids through each other show that there may be a motion of the small parts of a body which is not perceptible to us."

The small parts are not assumed to be of one uniform magnitude, nor even to have magnitude and figure at all. Each of them must have mass, and they must have the power of acting on each other, when near enough, like visible bodies. The properties of body are determined by the configuration and movement of its small parts.

The investigations which lead to the conclusions arrived at in this branch of mechanics are chiefly based upon the nature and action of gases under what is known as the 'kinetic theory of gases.' It is, in brief, that a gas consists of a swarm of perfectly elastic molecules in constant motion with different velocities, acting on each other only when they come infinitesimally close, — a thing, however, which rarely happens, — at all other times moving freely in unobstructed paths, the dimensions of the unoccupied spaces being immensely great compared with the diameter of the molecules. The molecules impinging against the sides of the containing vessel are driven back in all possible directions, and without loss of velocity.

We have nothing to do here with the theory, as such, nor can we follow, even in outline, the mathematical and experimental reasoning in its favor. Proposed by Kroenig, given form to by Clausius, modified and sustained by Maxwell and Thompson, it holds the favor of the scientific world. It is directly dependent upon the theory of molecules, and perhaps now on Sir William Thompson's theory of vortical atoms.

Let us see what a molecule must be in order to accommodate itself to the work required of it by the physicists. A molecule is a system composed of atoms, — the atom being the ultimate mass-unit. Molecules are complex and of different kinds, the several orders depending upon the number and arrangement of the atoms which compose them. This combination is not permanent, but may be broken up from one cause or another; in which case new molecular combinations result, and the nature of the substance which they compose is changed. This is seen constantly in chemical reactions, as in the case of water; the particles are the result of the aggregations of molecules of a definite composition; when this molecular condition is broken up, as may easily be, there appear two gases, one composed of oxygen molecules, and the other of hydrogen molecules. A system of atoms, then, which hang together for a measurable period, or until violently disrupted, is a molecule.

The dimensions and weights of molecules have been calculated with a sufficient degree of certainty to carry conviction to the scientific mind, though the demand upon the credulity of a layman is such as to make him doubt if he be not dreaming. Clerk Maxwell states that the fair probability is that 200,000,000,000,000,000,000 molecules of hydrogen would weigh a milligramme. The diameters of these miniature worlds are on much the same scale of magnificent littleness. About 2,000,000 molecules of hydrogen in a row would measure a millimeter. Professor Maxwell tells us that Loschmidt illustrates these measurements by the smallest possible magnitude visible to the microscope. A cube whose edge is the four-thousandth of a millimeter may be taken as the *minimum visibile*. Such a cube would contain from sixty to one hundred million molecules of oxygen or nitrogen; but since the molecule of organized substances contains, on an average, about fifty of the more elementary atoms, we may assume that the smallest organized particle visible under the microscope contains about two million molecules of organic matter. Another way of putting it — for it passes all real comprehension in any form — is: if one were to attempt to count the number of molecules in (say) a metallic mass the size of a pin's head, allowing the count to be at the rate of a thousand a second, it would require about two billion five hundred million years! And yet, the space occupied is by no means full — would be called rather empty.

Professor Tait, in his "Recent Advances," gives his readers a much needed preliminary caution in entering upon the subject of the structure of matter. He reminds them of what "every one worthy the name of mathematician" must know, "that there is no such thing as absolute size, — there is relative greatness and smallness, — nothing more. To human beings, things appear small which are just visible to the naked eye — very small when they require a powerful microscope to render them visible. The distance to a fixed star from us is enormous compared with that of the sun; but there is absolutely nothing to show that even a portion of matter which to our most powerful microscopes appears as hopelessly minute, as the most distant star appears in our telescopes, may not be as astonishingly complex in its structure as is that star itself, even if it far exceed our own sun in magnitude. Nothing is more preposterously unscientific than to assert (as is constantly done by the quasi-scientific writers of the day) that with the utmost strides attempted by science we should necessarily be sensibly nearer to a conception of the ultimate nature of matter. Only sheer ignorance could assert that there is any limit to the amount of information which human beings may in time acquire of the constitution of matter. However far we may go, there will still appear before us something further to be assailed. The small separate particles of a gas are each, no doubt, less complex in structure than the whole visible universe; but the comparison is a comparison of *two infinities.* Think of this and eschew popular science, whose dicta are pernicious just as they are the outcome of presumptuous ignorance."

This reminds one of Pascal's reflection that man is placed between two infinities, the infinitely great and the infinitely little, the one just as marvellous as the other. It is natural that we should feel, when an object passes out of reach of the senses that it loses those clearly marked differences which distinguish it while within their scope; though perhaps the mind, at least in our day, is more ready to delve down into the unknown in the direction of the infinitely little, than to ascend into the realm of the infinitely great. Clerk Maxwell in speaking of this minute world points out that we are set face to face with physiological theories, and warns the histologist not to imagine that structural details of infinitely small dimensions can furnish an explanation of the infinite variety which exists in the properties and functions of the most minute organisms. He remarks that while one microscopic germ is capable of developing into a highly organized animal, another germ, equally microscopic, becomes, when developed, an animal of a totally different kind. "Do all the differences, infinite in number, which distinguish the one animal from the other," he asks, "arise each from some difference in the structure of the respective germs? Even if we admit this as possible, we shall be called upon by the advocates of Pangenesis to admit still greater marvels. For the microscopic germ, according to this theory, is no mere individual, but a representative body, containing numbers collected from every rank of a long-drawn ramification of the ancestral tree, the numbers of these members being amply sufficient not only to furnish the hereditary characteristics of every organ of the body, and every habit of the animal from birth to death, but also to afford a stock of latent gemmules to be passed on in an inactive state from germ to germ, until at last the ancestral peculiarity which it represents is revived in some remote descendant.

"Some of the exponents of this theory of heredity have attempted to elude the difficulty of placing a whole world of wonders within a body so small and so devoid of visible structure as a germ, by using the phrase 'structureless germs.' Now, one material system can differ from another only in the configuration and motion which it has at a given instant. To explain differences of function and development of a germ without assuming differences of structure is, therefore, to admit that the properties of a germ are not those of a purely material system."

Resuming our consideration of the molecule, we are told that the results of spectroscopic investigations confirm the fact, already arrived at, that in gases, when in a rarefied condition, each molecule is at such distances from every other that it executes its vibrations in an undisturbed and regular manner.

But serious difficulties present themselves in the way of the unit-atoms out of which the molecule is built. To quote our authority again: "The small, hard body imagined by Lucretius, and adopted by Newton, was intended for the express purpose of accounting for the permanence of the properties of bodies. But it fails to account for the vibrations of a molecule as revealed by the spectroscope. We may indeed suppose the atom elastic, but this is to endow it with the very property for the explanation of which, as exhibited in aggregate bodies, the atomic constitution was originally assumed. The massive centres of force imagined by Boscovich may have more to recommend them to the mathematician, who has no scruple in supposing them to be invested with the power of attracting and repelling according to any law of the distance which it may please him to assign. Such centres of force are no doubt in their own nature indivisible, but then they are also, singly, incapable of vibration. To obtain vibration we must imagine molecules consisting of many such centres; but, in so doing, the possibility of these centres being separated altogether is again introduced. Besides, it is questionable scientific taste, after using atoms so freely to get rid of forces acting at sensible distances, to make the whole function of the atoms an action at insensible distances."

To meet these difficulties, Sir William Thompson, taking up the work of Helmholtz on the subject of vortical motion — work based upon the foundations laid by Lagrange and other great mathematicians of the last century — has advanced a new phase of the composition of matter. It is called the theory of Vortex-atoms, and is received with decided favor by the mathematicians. He would be a bold man who should say that he fully understands it. It relies entirely upon the higher analysis in method — indeed its author and those who pretend to follow him, complain that the calculus and other known methods have not sufficient grasp to handle the questions it presents. We therefore cannot be expected to do much to make it comprehensible; but we may get some notion of its general trend by sticking pretty close to Professor Tait's explanation.

First, it is easy enough to understand the vortex-ring. We have an illustration in the rings an expert smoker sometimes makes for the amusement of children — making a round hole with his lips and emitting the smoke in puffs. Indeed, it seems to have been an illustration of this kind, made by Professor Tait (only he used a box filled with ammoniacal gas, and driven out of a hole in one end by blows upon a flexible substance at the other), which first suggested the theory to Sir William Thompson. Such a vortex-ring moves as if it were an independent solid; and " if the air were a perfect fluid, — if there were no such thing as fluid-friction in air," — such a vortex-ring would move on forever as a solid mass. When two such rings, even in air so little complying with the conditions of a perfect fluid "impinge upon one another, they behave like solid elastic rings. They vibrate vehemently after the shock, just as if they were solid rings of india-rubber." These rings may be varied — practically, to some extent — theoretically, without limit. Vortex-filaments may exist with any number of knots and twists in them — and, practically they may be made to link together and present a certain degree of permanence.

All this is yet worlds away from the vortex-atom; but assume a perfect fluid, possessed of inertia, invariable density, and perfect mobility, — call in the powerful aid of mathematical analysis, — refine and continue to refine, until the ordinary thinker has been hopelessly left behind, and somewhere in such a thought-spun world the vortex-atom may exist.

If the proposed theory be a true explanation of the facts of nature, then rotary motion of the proposed perfect fluid (which itself is not matter; matter being a mode of motion of this fluid) is the mechanical basis of all that appeals to our senses. The theory, however, is not yet established; but we cannot fail to see in the persistent efforts to overcome the difficulties in the way of the atom, how desperately shaken the scientific world is as to the nature of matter. We shall have to wait patiently on the mathematicians (the controversy being far beyond all mere experimental method), since, as Professor Tait tells us, "to investigate what takes place when one circular vortex-atom impinges upon another, and the whole motion is not symmetrical about an axis, is a task which may employ perhaps the lifetimes, for the next two or three generations, of the best mathematicians in Europe; unless, in the meantime, some mathematical method, enormously more powerful than anything we at present have, be devised for the purpose of solving this special problem."

That there are serious difficulties lying in the way of the vortex-atom is freely confessed. Should it succeed in getting itself thoroughly established from the rotarial and kinematic side, it has still grave difficulties to encounter. Among these is, the demand upon it, as upon all other theories, to account for mass and gravitation. In Thompson's theory, mass — that is to say, inertia — is assumed in the perfect fluid. But inertia is a property of matter, and not a property of motion. The difficulty remains without serious attempt at solution.

With regard to gravitation, the mathematicians agree that present methods are sufficient to show that the vortex-atoms cannot be allowed to exert a pull on each other such as is now assumed to exist between bodies; so that a serious alternative is presented at once — either the theory must go, or this assumed attraction of gravitation is not a fact of nature.

This looks threatening, but there is a possibility of escape. Le Sage, of Geneva, put forth a theory, now nearly a hundred years ago, to explain gravitation — or rather, to explain it away, so far at least as the notion of a pull between bodies enters it. It is entirely in line with the trend of modern mechanics, and serves the turn of Thompson's theory fairly well; and is also far enough out of the run of common thought to delight the wildest scientific imagination. A brief statement must suffice.

Le Sage proposed to abandon the notion that bodies or masses, large or small, are pulled together by solicitations subsisting between material elements, and to adopt instead the notion that they are driven together by the impact of streams of ultra-mundane corpuscles which are flying about in all possible directions. These he supposes to come with enormous velocities from regions infinitely remote, and to be so minute, compared with the distances between them, that a collision can rarely happen. If a particle of gross or common matter could be exposed to this ceaseless pelting from every conceivable quarter, the result would be that the blows would, on the whole, neutralize each other, and the particle would retain its place; but this could never happen, since the other particles in existence must be taken into account. Every particle must shield every other from a portion of this pelting, and just in so far each particle will have its equilibrium disturbed, and so be driven in the direction of least resistance, with a velocity precisely the same as that now effected by the supposed pulling power of gravitation, the law of the inverse squares obtaining as before. Of course the space occupied by gross matter must be so little filled by the essential atoms composing it, as to permit the enormously greater number of ultra-mundane corpuscules to pass through unimpeded.

By this theory the vortex-atoms would be saved the need of being what is called heavy, or of imparting weight to the bodies they compose; all this would be relegated to the ultra-mundane — ultra-stellar — corpuscules, without inquiring as to what that region can really be, or how the projectile energy is originally generated.

There is another fundamental hypothesis of physics which demands attention, and that is the assumption of the existence of an inter-stellar medium, called ether. The marvels of the atomic theories have drawn heavily upon our imaginations, but in this a still larger demand is made upon us.

The need of something to fill the immeasurable spaces between the myriad worlds is, from a metaphysical point of view, very urgent. To say there is 'nothing' is to use the word in a sense far this side of the absolute naught. To say that it is filled by space is, as we shall see further on, misleading and void of meaning. But we pass this.

The need of some sort of medium from the physical standpoint is equally urgent. The question presents itself sharply in the undulatory theory of light. The Newtonian or Corpuscular theory was — for it may be now considered dead — that light was a material substance, consisting of extremely small corpuscules emitted from luminous objects and producing the sensation of light by impinging upon the eye. They were supposed to be shot to the earth from the sun and other luminous bodies, and thus the demand for an intervening substance was obviated, at least, from the physical point of view; but difficulties of its own immediately presented themselves. The theory was questioned, almost from the beginning. With the enormous velocity, — nearly two hundred thousand miles a second, — and with the least possible mass in the corpuscules, the momentum would inevitably destroy such a delicate organ as the eye; but the most refined test fails to detect the slightest possible signs of impact. There are many other difficulties in the way of the emission theory, the most decisive perhaps being found in its inability to explain many phenomena, especially those due to what is now known as 'interference.' As a single example, in the case of light from one definite source, divided into two portions by mirrors, so that they shall fall on a screen in such wise that the paths passed over shall differ by a certain very slight distance, while either part alone produces light only, when the other part is added, dark spaces appear. There is no possible reason for these dark areas except the added light; so that the increase of light produces darkness. This could not be if it were a substance.

The theory of Newton has given way to what is known as the Undulatory, or Wave Theory of light. It was originally suggested, perhaps, by Hooke, a celebrated contemporary of Newton, but it first took shape in the hands of the great astronomer and mathematician, Christian Huygens. It is not necessary to say more than that, granted an intervening substance of high elasticity, a disturbance at one point must be propagated with great rapidity throughout its extent, and without the consequences which would attend a projectile under like velocity.

Such media, under the name of ethers, had been freely assumed in the past, extending back to the remotest antiquity, to account for all manner of phenomena. At the time, and shortly after Huygens proposed his luminiferous ether, the feeling in the minds of scientific men against ether hypotheses, no doubt, acted to prejudice most students of science against the new theory. Certain it is, that it was opposed by the most eminent mathematicians and physicists, Laplace, Malus, Biot, Brewster, etc., and that the theory of Newton maintained its ground until, in the early part of this century, it was attacked by Dr. Thomas Young, backed by Augustin Fresnel. The contest was long and animated, but they succeeded in adducing so many phenomena which the corpuscular theory could not account for, and which the other handled easily, that the old theory had to give way. It was given its *coup de grace*, perhaps, by Foucault when he succeeded experimentally in proving that light travels more slowly through water than through air, — a condition of things demanded by the new theory, but directly in contravention of the requirements of the old.

The wave theory has not only met the demands upon it in explaining the observed phenomena of light, but years ago it enabled Fresnel to anticipate 'circular polarization,' and recently the theory, in the hands of Sir W. Rowan Hamilton, achieved a triumph in optics quite analogous to the astronomical feat of Leverrier and Adams in discovering the planet Neptune.

The vindication of the existence of a luminiferous ether seems complete; so much so, at least, that all cognate phenomena have come to lean confidently upon it for the ground of their explanation. But the demands of the theory upon the nature of the medium are exceedingly exacting. The phenomena of polarization, as explained under the theory, make it absolutely necessary that the vibration shall be transverse to the direction of propagation, and there is thus imposed upon the ether the necessity of being a solid. But this is not the worst; it must be a solid of such rigidity as passes all comprehension. Its elastic force is reckoned to be over one million times that of air at the surface of the earth, which so far exceeds that of any known substance as not to be named in the same breath: a pressure which may be represented by the weight of a pile of granite blocks a foot square at the base and something like twenty million miles high. And yet we, the earth, and all moving bodies, dense and rare, pass through it, or it through them, without discoverable resistance.

It is still an open question as to whether this ether is continuous or discrete. It is not ordinary or gross matter, but whether it is molecular in structure, or what is its constitution, is not agreed upon by science. Neither is it perfectly homogeneous, *i.e.* equally distributed in all kinds of solids and other forms of gross matter, since light travels at different rates through substances of different densities, which means that the elasticity of the medium of transmission must differ in such substances. The theory of light, assuming the existence of the luminiferous ether, has still difficulties of its own to meet, and the demand is rapidly returning for a variety of ethers. At any rate, the mathematicians show little disposition to settle down permanently upon this single hypothesis, or to agree as to the demands to be made upon it.

But, not to pursue the subject further, it must be admitted that the hope of reducing all phenomena to a homogeneous substratum free from contradictions is very remote; and even if this could be accomplished, the old world-cry would still go up, What is it? and Whence came it? Further than this, it will hardly do for the physicists to cast scorn upon the metaphysicians on account of the refinements and subtleties in which they sometimes indulge, in view of the purely metaphysical regions and the metaphysical speculations into which science is itself so necessarily driven.

As it is, we may safely conclude that the physicists do not know what 'thing' is more certainly than the metaphysicians. Even if the conclusion be that atoms are definite material entities or world-stuff, the question is only pushed back to one deeper still. What and why they are at all, rises up and demands an answer as surely as any phenomenon in nature. Clerk Maxwell says: "In the present state of science, we have strong reasons for believing that in a molecule, or if not in a molecule, in one of its constituent atoms, we have something which has existed either from eternity or at least from times anterior to the existing order of nature. But besides this atom, there are immense numbers of other atoms of the same kind, and the constants of each of these atoms are incapable of adjustment by any process now in action. Each is physically independent of all the others.

"Whether or not the conception of a multitude of beings existing from all eternity is in itself self-contradictory, the conception becomes palpably absurd when we attribute a relation of quantitive equality to all these beings. We are then forced to look beyond them to some common cause or common origin to explain why this singular relation of equality exists, rather than any one of the infinite number of possible relations of inequality."

Professor Maxwell refers to Sir John Herschel's remark that atoms are to be compared to 'manufactured articles' on account of their uniformity, and, after giving the several possible meanings the expression may bear, says: "Which of these was present to the mind of Sir John Herschel we cannot now positively affirm, but it was at least as likely to have been the last as the first, though it seems more probable that he meant to assert that a number of exactly similar things cannot be each of them eternal and self-existent, and must therefore have been made, and that he used the phrase 'manufactured articles' to suggest the idea of their being made in great numbers."

CHAPTER XXV.

MATHEMATICS NOT ULTIMATELY EXACT.

Position of mathematics in scientific inquiries. Mathematical processes develop contradictions. Surds. Asymptotes. Graphical illustration. Cissoid of Diocles. Other cases. Right-lines intersecting with no common point. The concept 'infinity.' Illustration.

SINCE the days of Newton, the position of the mathematicians with respect to physics has been assured; but it has not been until our own times that their supremacy in this regard has been fully recognized. They absolutely dominate mechanics, and mechanics is now admitted to be the very soul of physics. After mathematical analysis has laid hold upon a subject, and pronounced a conclusion, the case is closed, in so far, at least, as the facts assumed are warranted. In the hands of genius this inexorable thought-machine has accomplished wonders, and the mathematician still has wide fields before him, with, no doubt, better methods yet to come. We hear Professor Tait crying out for a more powerful instrument with which to handle successfully the vortex-atom; and it is not unreasonable to expect, in view of the recent splendid discovery of Quaternions, that other and still more subtle analytical methods will be brought to the aid of physics.

Recognizing the title of the mathematics to the front rank of scientific thought, and with an admiration little short of reverence for its magic powers, it will not, I hope, be thought out of place to show, in an elementary way, that not only are there serious incompatibilities in the subject-matter submitted to investigation at its hands, but that the instrument itself, though the most perfect known, has its own incompatibilities — a fact which shows how impossible it is to raise even the most exact methods of thought above the possibility of emerging contradictions.

In the first place, Arithmetic is full of incompatibilities. As one out of any number of examples, take the case of a surd. The $\sqrt{7}$ (say) must have a definite value, and yet it has not, — or rather its value is neither entire nor fractional. We can approximate it as near as we please, thus, 2.645751+, and the decimal can be extended indefinitely; but if it were drawn out across the earth's orbit to Aldebaran, there would yet be something lacking. But, it may be said, since this difference is growing less and less at each remove, it would disappear at last if the operation could be extended far enough. Not so; it is easily shown[1] that the exact value is impossible. Thus it appears that while this value must be somewhere between 2 and 3, it cannot be at any exact point between, which is a flat contradiction.

The incommensurability of magnitudes discovers itself everywhere in geometry, as in the rectification and quadrature of the circle, the side and diagonal of a square, etc.

Imaginary expressions, that is, the indicated even roots of negative quantities, are also inconceivable. It is impossible that any such root can exist; and yet the Algebra has no difficulty in handling such expressions — combining and recombining them and arriving at perfectly definite and correct results.

The asymptotic curves are also examples of incompatible

[1] The proof is very simple. Thus, let the $\sqrt{a}$ be any surd. If its value be expressible by a fraction, let $\frac{p}{q}$ be such fraction in its simplest form. Then $\sqrt{a} = \frac{p}{q}$, and from this we have $a = \frac{p^2}{q^2}$. But $\frac{p^2}{q^2}$ must be an irreducible fraction; so that we have an entire quantity equal to an irreducible fraction, which is absurd.

conceptions. That a curve shall continue to approach a line and never reach it, is the characteristic of all asymptotes. This contradiction can be made obvious without the aid of mathematical analysis; and there are so many people who declare that they will not believe what is contradictory, that it may not be amiss to insert it.

Imagine a bit of cardboard, rather large, set up vertically, as shown in this figure.

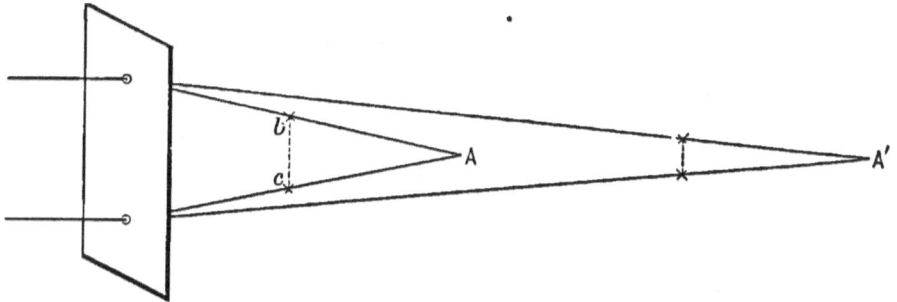

Let there be two holes, one near the upper and the other the lower edge. Pass cords through these holes and unite the ends on the right as at A. From this point A mark a point on each cord, b and c, exactly at the same distance from A. Now keeping the cord stretched, let the point A move to the right, always at the same distance from the floor. It will be seen at once that the marked points b and c must continually approach each other as A moves to the right, and that theoretically they never can come quite together; so that, though the two points approach each other forever, they never can meet. A horizontal line through AA' would be the common asymptote to the loci of the two moving points.

Again, it seems plain enough to common sense that no exact area can be enclosed so long as the bounding lines do not absolutely close on each other; and yet, the mathematics presents many cases in which areas are definitely calculated, though the limiting lines do not, and never can meet. Take as an example the Cissoid of Diocles. The curve may be described by points as follows:

Let AB be the diameter of a given circle, and from the extremity A of the diameter draw a straight line AC to meet the tangent line at the other extremity, and mark the point P at a distance from C equal to the intercepted cord AD. Determine any number of other points in like way on other lines through A above and below the diameter AB. The locus of all the points so found will be the Cissoid. It manifestly has two infinite branches, and the tangent line will be a common asymptote. Now, although by the conditions, the extremities of the two branches can never reach the tangent, nevertheless it is demonstrable that the area between the tangent and curves is exactly equal to three times the area of the circle.

The following is a still more remarkable case. This curve, whose equation is given in the footnote on the following page, can never touch the axes X and Y, or, which is the same thing, constantly approaches the

MATHEMATICS NOT ULTIMATELY EXACT.

axes above and below, and reaches them at infinity.[1] The area comprehended between the curve and the axis of Y above the line CD is proved to be exactly equal to the square $CDEO$, while the area of the portion $XEDB$ is infinite. Since the curve continually approaches the axis of X, it is impossible to see why it does not as certainly close on X, as the upper end does on Y; or how the excluded portion of the plane can ever get within, so as to enable the area to be infinite.

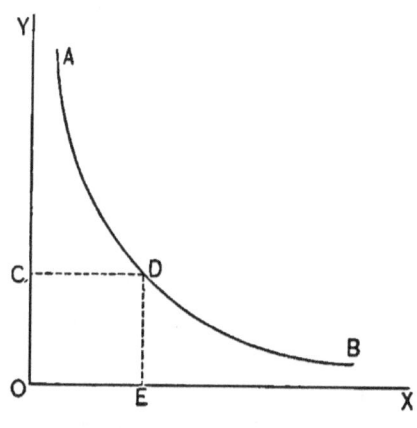

But, passing over innumerable contradictions which show themselves in the divers 'orders of contact,' 'singular solutions,' and indeed throughout the Infinitesimal Calculus, let us take a case of dire contradiction, as, for example, the proposition that two straight lines in the same plane and inclined to

---

[1] The equation of the above curve is $y^2 = \dfrac{1}{x}$.

Differentiating, we have $dx = -\dfrac{2\,dy}{y^3}$, whence

$$s = -\int \dfrac{2\,dy}{y^2} = \dfrac{2}{y} + c.$$

Estimating the area from the axis of Y, giving $y = \infty$, we have

$$0 = \dfrac{2}{\infty} + c, \quad c = 0, \quad s' = \dfrac{2}{y}.$$

Making $y = 1 = DE$, we have $s'' = 2 = ADEOY$; that is, twice the square $CDEO$. Taking the area between the limits $y = 1$ and $y = 0$, we have

$$\text{Area } BDEX = \dfrac{2}{0} - 2 = \infty.$$

each other, do, and do not, intersect in a point. Take two lines, given by their equations, such that upon combining, the co-ordinates of their common point prove to be surds. Now, as we have seen, exact values for these co-ordinates are impossible, and therefore there cannot be a definite point which answers to them; and since the lines must intersect in an exact point, they cannot logically intersect at all.

This is all fine spun, of course, but it shows that the difficulties of thought which show themselves in studying the composition of matter, time, space, cause, and in the whole domain of metaphysic, discover themselves also in mathematics, — that while in this science exactness is the rule, absolute determinism cannot be predicated of the exactest of all the exact sciences.

These difficulties no more shake the foundations of mathematics than the insoluble contradictions in metaphysic disturb the current of every-day thinking; but they do show that logic, even in its purest form, is not an infallible instrument, but has its 'little rifts' which only the more surely bespeak it kin to the universe of mystery. One of the great evils of the day is that a large, and that a vigorous-minded, class of men, have undertaken to smooth out the folds in the world's vesture, and, with the rush and whirl of an unbalanced science, trample out of the heart of man all sense of wonder — all thought of the Ineffable which moves him to be silent and adore.

Before leaving this subject, it will be well to consider for a moment how the 'infinite' is to be construed. The word is used very loosely on all hands, but even in a technical and scientific sense there is often confusion of thought.

To most minds infinity means something very great, either in length, or volume, or power, or minuteness, but still carrying with it some unit of comparison by which to estimate it. While in a popular way the word serves a good purpose in this sense, this meaning cannot be allowed from a scientific point of view.

The infinite is not a huge finite; indeed, it is just what the finite is not. It is therefore a great mistake to suppose that one can follow (say) a line on and on, and so arrive at infinity. The infinite absolutely refuses to be composed or made up by adding more and more, or by any process of reduplication. It cannot be *approached*—it is only to be reached — (if that word can be used at all in this sense) —*per saltum;* and the leap is just as great last as first. For example, suppose we take the distance to the remotest fixed star as a unit — a distance (say) requiring light a thousand years (or any other number one likes) to travel; and suppose that every millionth of an inch in it were suddenly expanded into a length equal to the whole, and all strung out together, it would still be just as far from infinite length as any one of those unexpanded fractions. Not the most infinitesimal step has been taken to pass a chasm which is not even reached until the finite has ceased to be. The infinite differs from the finite, not in degree, but in kind. It is 'other' to it, and in final analysis cannot be permitted to retain the remotest likeness.

This may be illustrated very simply by the tangential function of an arc. The tangent grows at first with moderate slowness from the origin, but with constantly increasing strides for equal increments of arc, until, when the arc approaches 90°, the length of the strides becomes each inconceivably greater than the last — greater than all the distance travelled from the origin — until, when the secant becomes nearly perpendicular to the initial diameter, the mind sinks under the overpowering sense of immensity; but the arc is not yet 90°, — the mad flight goes on, and it is not until the last minutest difference of a difference of arc is passed — itself an impossible conception, — not until the secant lets go its touch upon the tangent line and becomes parallel to it, that the function suddenly leaps the chasm which separates it from all finite values behind, and *is* — not *becomes* — infinite. It is not that in the last infinitesimal of arc added, there is added to the stupendous length already gained by the tangent, enough to make it infinite, — for all that is zero in the presence of infinity, — but that a world-wide change has taken place, and it has ceased to be what it was — the finite — and become what it was not — the infinite. If one rightly takes this in, one gains some faint notion of the incommensurable character of the infinite; and yet, there is no point in the whole movement in which the same incompatibility of thought does not present itself. It is the same marvellous truth which has met us all along our way — the passage from the finite to the infinite and back again — phenomenon of change in any sort — incomprehensible in thought, and yet ever taking place in fact.

CHAPTER XXVI.

THE METAPHYSICAL ATTITUDE OF CHANGE. CAUSE.

The problem of Change. Quotation from Plato. The problem of Causation. Influence 'passing over.' Doctrine of 'Occasionalism.' 'Pre-established Harmony.' 'Divine Assistance.' Lotze quoted. Causation, as such, inexplicable.

WE go on to show the incompatibilities of thought which discover themselves in the conception of 'change,' of 'motion,' and of 'cause.' Changes are so natural and so necessary that it seems to most minds a mere impertinence to inquire what 'change' is. In practical life there is no need to ask such a question. Everybody knows what is meant, so long as no explanation is required, and this practical knowledge would do perfectly well if one could keep the question out of mind. But that is impossible, at least for those who are not content with the mere 'bread-and-butter' side of existence. The question is asked, and has been asked as far back as the dawn of speculative thought; and it still waits such an answer as shall carry conviction to those who rightly take in the difficulty.

The problem now is not, How can change be? for that would simply throw us back upon the question which lies behind all inquiries — How can anything be? The question is, How can we account for the mutations we see going on, so as to make "the total idea of it without contradictions, and adequate to those facts of experience which we wish to designate by means of it"?

Take, for example, a body at rest: How can it become a body in motion? Being at rest, and having no motion, by hypothesis, in moving, must it not move while it is at rest? It cannot wait until it has motion to move, since it would in that case wait forever. It must have motion, then, while it has it not, — a contradiction sufficiently glaring. The same is true of a body in motion: How can it ever come to rest? It cannot be at rest so long as it is moving; that is, up to, and into the point just before that at which it is to rest; and from this, it must move to the next and last. It must carry its motion, then, to the very end, and motion and rest again come together. The difficulty is the same at any point of its path where there is a change of velocity. Since the velocity changes, the change must be at some point, and the body must arrive at the point with one velocity, and leave it with another; that is, have two velocities at the same time and at the same point.

The infinite variety of forms in which this difficulty of thought can be thrown may be seen most thoroughly and beautifully in the Dialogues of Plato. I quote a few sentences from the "Parmenides" (Jowett), which is taken up with the discussion of the 'One and the Many.'

"Further consider, whether that which is of such a nature can have either rest or motion.

"Why not?

"Why, because motion is either motion in a place or change in self; these are the only kinds of motion

"Yes.

"And the one, when changed in itself, cannot possibly be any longer one.

"It cannot.

"And therefore cannot experience this sort of motion?

"Certainly not.

"Can the motion of one, then, be in place?

"Perhaps.

"But if one moved in place, must it not either move round and round in the same place, or from one place to another?

"Certainly.

"And that which moves round and round in the same place must go round upon a centre; and that which goes round upon a centre must have other parts which move round the centre; but that which has no centre and no parts cannot possibly be carried round upon a centre?

"Impossible.

"But perhaps the motion of the one consists in going from one place to another?

"Perhaps so, if it moves at all.

"And have we not already shown that one cannot be in anything?

"Yes.

"And still greater is the impossibility of one coming into being in anything?

"I do not see how that is.

"Why, because anything which comes into being in anything, cannot as yet be in that other thing, while still coming into being, nor remain entirely out of it, if already coming into being in it.

"Certainly.

"And, therefore, whatever comes into being in another must have parts, and the one part may be in that other, and the other part out of it; but that which has no parts cannot possibly be at the same time a whole, which is either within or without anything."

This is not a mere juggle of words, as one might be at first tempted to declare. It is a difficulty of thought which we have so often encountered, and shall be thrown back upon constantly as we proceed. It is of the same nature as the difficulty which the physicists find in the construction of matter;

and discovers itself whenever thought attempts to transcend the sphere of relation, and deal with the ultimate in any direction.

Granting, what we can neither deny nor explain, that change is an actual factor in the economy of nature, the further inquiry arises, How does change come about? The two questions are clearly separable: one is of the 'What,' and the other is of the 'How.' One thing acts upon another. We see the change which takes place, and, passing the question as to what the essential difference is between the new and old order of states or conditions, we want to know now in what way one thing acts upon another to produce the new. This is the problem of 'cause,' of which we have already had somewhat to say, but which requires some further consideration. We shall follow Lotze in our outline.

In the first place, the proposition, Everything has a cause, is not quite correct. In mathematical truth we do not seek a cause. When we say a triangle has three sides, we can give no reason, and we do not look for any. Neither do we look for cause in the actual, so long as it simply abides. It is only when *change* takes place, or is conceived to be happening, that we want to know the reason. "The 'Being' of an existence can in itself be regarded as perfectly unconditioned and eternal. It is only the special nature of what exists that can, on manifold other grounds, excite a doubt respecting its unconditioned existence and inquiry after its origin. Even such an investigation, however, must terminate in the recognition of some unconditioned being or other."

In an *effect*, we may distinguish the content and the actualization. The 'content' is that which distinguishes the event from other events, as when a spark is applied to gunpowder, the effect is known to be an explosion instead of (say) a change of color or a change of substance. The 'actualization' is the happening or act of changing without regard to the particular result after the change is complete. The difficulty lies chiefly in the actualization; how does one thing act on another?

The common answer is that an 'influence ... passes over' from one element to the other (*causa transiens, influxus physicus*), and produces the effect. This seems to have a meaning at first look, but upon examination it will be found quite empty.

In the first place, what is it that passes over? If we think of it as something real, — a constituent part or element, c, which separates itself from A, and, moving over, unites with B, this is quite comprehensible as a fact, but it is not what is meant by 'cause' in any right sense. It is but the translation in detail of A to, or into, B. "When water (c), for example, with all its properties passes over from A to B, the only effect is that those properties now appear at the place B (which becomes moist), and vanish at the place A (which dries off)."

The only other way in which we can regard this passing over is that something not real, but belonging to A as a potency, attribute, or condition, shall proceed out of the thing which affects, to the thing affected; but such a potency, whether it be called 'state,' 'influence,' 'efficiency,' 'force,' or what not, so long as it is not a part or constituent of A, is but an attribute and cannot exist apart from, and independent of, its subject. A 'state' and the like can never be set loose from the 'thing' A so as to exist by *itself* for an instant between A and B, unsupported by a subject, and then attach itself to B. In other words, there is a gulf fixed between cause and effect, narrow as it may seem, across which no mere force or influence — nothing — can pass which is not itself 'thing'; and if it be 'thing' we have a mere translation, and no true effect.

But even if it could be made comprehensible how this gulf could be passed, this would but bring the compelling power near to, or against, the thing to be moved; and the real question then emerges, Why is the vicinity of this something to the thing to be moved of such importance to it that it must move? That is, What causes the thing B to move? We have not really advanced a step. The causal nexus is as entirely undiscoverable as at the first.

This is a point which most minds do not apprehend readily. Let us take an illustration.

A push or pull on a rod is communicated from one end to the other. The molecular springs in the first film moved react upon those of the next, and so on to the end. Now it is easy enough to say that the second and consecutive films must move, but this is just the question: Why must they move? If one says, "Because they must," the answer is safe, and in reality there is no other to make; but this is a surrender of the point. It is not at all like a conclusion which follows apodictically from logical premises, for there the contradictory is unthinkable. In this case it is not at all so. Here the conclusion is clearly based upon experience. If one body, in impinging upon another, had not constantly produced before our eyes this resultant change in the body struck, there would be no expectation of a change in consequence of the collision; and as it is, we can without difficulty conceive of such a sudden alteration in the physical character of bodies that no such result would take place. That we cannot do in the case of a necessary deduction in a logical process, unless we conceive of a change, not in the subject-matter but in the powers of thinking. There is thus a manifest difference.

All this has been so clearly seen by the great thinkers of the world that a number of schemes have been proposed to escape the difficulty. Let us look at some of them very briefly.

First, the Cartesian doctrine of Occasionalism, as developed especially by Guelincx, had for its especial object to explain how mind and matter act and react upon each other under the dualistic postulate of Descartes, that mind and matter are totally unlike, — as much so as if they belonged to wholly different worlds. It proposed to abandon all thought of the direct action of one thing on another, and to consider the world, in its succession of events, as utterly unconnected by any causal nexus, holding all antecedent states or conditions to be but the occasions or signals upon which the effects follow; the true compelling cause residing in Deity.

Manifestly the difficulty is not thus removed, but recurs in a new form, weighted down with other difficulties of its own. The question now is, How does the thing affected *know*, and how is it made to respond to, the signal given? The answer is as impossible as in the case of a supposed influence 'passing over.' Its own difficulties are that it is contrary to the convictions of experience, and introduces all manner of unrealities for what at least seems natural.

A second theory is what is known as 'Pre-established Harmony.' This theory, invented by Leibnitz, assumes that the entire world, spiritual and physical, down to the minutest details, was pre-arranged by the Author of Creation, so that each and every event follows its precedent so-called cause with unfailing certainty — not because there is any dependence one on the other, but because they were so arranged as to happen in an established sequence. Thus, when one wills to raise one's arm, the muscles and the whole physical organism are moved, not by the will or any self-energy, but because it was so ordered that upon the particular volition happening, that particular movement should take place.

Even granting that the world was so ordered in the beginning, when God withdraws himself from his work, what guarantee is there that the order so established will continue to subsist? If one thing does not act upon another, what would prevent the whole from falling into confusion; and how do we know that the world is going on as pre-arranged? It will be seen that the difficulty is not in the least avoided, but really made worse. There must be something constraining and compelling the orderly continuance of events, and this is the old question back again, with its own difficulties added. The famous illustration of the clocks, so constructed as to run exactly together, but without any possible connection or influence on each other, does not meet the case; for after the clocks are made and placed in perfect accord of movement, how shall either of them continue to carry out the purpose of the mechanician, if one wheel does not act upon and compel the movement of another? Thus the old difficulty steals back upon us.

Another form of the same theory presents the case, not as a fixed and previously determined order, but introduces a hypothetical element; God has not ordered all things absolutely in advance, but has established a provisional order, such that *if* a certain a comes to pass, a certain b will follow. This theory fails in the same manner as the categorical form just considered; for if a certain 'thing,' n, is compelled to pass into a certain state or condition, a, whenever another state or condition, b, happens to a second thing, m, then the n must take some notice of m's being present, and affected by b, before it can pass into the state a. The notion of causation is not escaped, since either m or b must have an effect upon n.

Still another form of the same general doctrine devised by the followers of Descartes, is called the theory of 'Divine Assistance'; and the especial point is, that while one thing cannot be the efficient cause of another, God by his own power compels the proper reaction which answers to the action. Even this theory does not get rid of the notion of causation, but contains it twice over. For, in order that God, in the light of our thought, may attach to every a its b, and to every c its d, it is necessary, first, that the presence of a or c shall have some effect on the Divine mind, and that the effect of one shall be different from that of the other; and second, it is necessary that God, in the order of his own laws, shall react upon the things in question; and so produce one effect in consequence of a, and another in response to b.

Lotze says, in summing up the results of the discussion of cause: "The conception of efficient Causation is inevitable for our apprehension of the World, and all attempts to deny the necessity of efficient Causation, and then still comprehend the course of the World, make shipwreck of themselves. But just as certain is it that the nature of efficient Causation is inexplicable; that is to say, it can never be shown in what way Causation in general is produced or comes to pass; on the contrary, all that can ever be shown is, what preparatory conditions, what relations between the real beings, must in every case be given, in order that this perpetually incomprehensible act of Causation may take place.

"That the inquiry into the 'bringing-to-pass' of efficient Causation is necessarily unanswerable, and in its very nature senseless, is shown by the *circulus* into which it straightway leads. For, if we want to get an insight into the causative process of Causation itself, we naturally take for granted, as something necessarily familiar, the causal efficiency of that very cause which is assumed to produce the Causation to be explained; we are therefore explaining efficient Causation by itself."

This does not in any wise affect our general conception of causation, and we must go on thinking of force or influence 'passing over' from one thing to another to produce effects. In this regard, causation is in no worse case than matter or mind, since we do not know what either mind or matter is. They are to us concepts, the actual content of which vary no doubt considerably for different persons.

It seems clear, however, that the assumption of independence or absolute separateness of 'things' must be abandoned. All efforts to bring isolated things into relation, once assuming them to be out, must utterly fail. In the case of a being A, affecting another, B, any state *a* which takes place in A, must for the very reason that it is in A *be* an 'affection' in B; but it does not necessarily *become* such an affection of B by means of an influence issuing from A.

"The foregoing requirements can be met only by the assumption that all individual things are substantially one; that is to say, they do not merely become combined subsequently by all manner of relations, each individual having previously been present as an independent existence; but from the very beginning onward they are only different modifications of one individual Being, which we propose to designate provisionally by the title of the Infinite, of the Absolute $= M$."

The formal consequences of this assumption, Lotze goes on to say in substance, are as follows: Any particular 'thing,' *a*, is but a special mode of the universal M, and any other definite thing, B, is another mode of M, etc. Every state which takes place in *a* is but a further differentiation of the *a*-mode of M and is therefore a state of M. From the nature of its own laws this modification must show itself in M, and it accordingly produces a further differentiation in some mode of M, say the *b*-mode. This modification thus appearing in *b*, seems to be and is the same thing as *a* acting on *b*.

Efficient Causation, therefore, is a necessary concept, and has no actual content; but, in an ultimate sense it is the Absolute itself; and the actuality is the result of self-activity somehow and somewhere. We can say one thing *really* acts upon another, provided we do not interpret the word 'really' to mean more than the word 'real' can mean when we say a 'thing' is real. We find here, as we have found before, and shall always find, that the ultimate in any case is known; but not by the understanding in its technical sense which always implies the relation of subject and predicate; but by that form of knowing which comes to us at first hand in the form of rational intuitions, and without which the understanding could have no ground. In referring it back to the One Infinite Being, the purpose is not to offer any theory of explanation of causation, as such; since 'the manner' in which it comes to pass, that even within the One Infinite Being, one state brings about another, remains still wholly unexplained; and on this point we must not deceive ourselves. How it is in general that 'Causal action' is produced, is as impossible to tell as how 'Being' is made. And yet, as a primordial fact, we know it certainly and in the same way as we know that we are what we are.

CHAPTER XXVII.

RELATION OF PERSONALITY TO SPACE AND TIME, MASS AND MOTION.

The concepts 'Space' and 'Time.' Subjective ground of Mass and Motion. Not self-subsisting realities. Find their reality in Personality. Reality of the cosmos personal. Soundness of scientific methods. No truth material. Personality necessary to truth. Personality not a phenomenon.

IT is not to be questioned that we have a necessary intuition of space, but it is seriously questioned as to whether space is an objective entity or not; that is, as to whether it belongs to the thing-world, or to the spirit-world. It can make small claim to be 'thing,' since it is incapable of affecting the sensibilities, or of being itself affected by anything whatever. The necessary condition of pure space is that it shall be empty, and therefore no-thing. It cannot even be permitted to enjoy the property or capacity of being a containing somewhat, though this is a firmly established prejudice, arising from our uniform experience of vessels, and other receptacles, for, upon removing the limiting surfaces, the notion falls away. This, however, does not at all affect our notion of distances from point to point in bodies, or between bodies, that is, of the trinal dimensions of extension; but these are clearly relations, and not entities. All possible objects must carry with them the notion of extension; so that the space-form must discover itself in any object. Whenever we think of air, or ether, or aught else of a dimensional character, we must have the notion of space; and there will always be lurking in the mind some 'thing' in the content of the space-concept. We cannot think of space as a *prius*, existing independently, and of things as afterwards made to fill it. As Lotze expresses it, things do not exist *in* space, but space exists *in* things.

In like way time has none of the characteristics of body, but is a presupposition of movement and change. Of eventless, empty time we can form no conception whatever; but just as body carries with it the necessary notion of extension, so movement carries with it the intuition of duration. Space is statical; time dynamical. Space is the ground and presupposition of mass, and time the condition and presupposition of motion, — the two fundamental postulates of mechanics, — and they both belong to the psychical side of personality.

With regard to Motion, we have the same sort of difficulties. What is Motion? An old Greek, it is said, was once asked this question by his scholars; he strode across the floor, and exclaimed, "There! you see it, but what it is I do not know." He was in no worse case than all the learned world has been in ever since. Ordinarily we accept what the old philosopher displayed to his scholars as motion, namely, 'change of place,' with respect to other things which appear fixed — 'relative motion.' This is comprehensible enough until we begin to ask what a fixed point is, and then we are thrown back upon the negation of motion for an answer. The circle is narrow enough — a point at rest is one not in motion; a point in motion is one not at rest; but this does not tell us what rest *is*, nor what motion *is*. The question is not, what is a moving object? but what is motion which we are made conscious of in the object? If it be replied that motion is *only* object in motion, then it follows that there is no such thing as motion *per se;* because it is *not*, when the object is at rest, and is in varying degrees as the body moves under an accelerating energy, gaining motion at every point, losing it under a retardation, until all is lost, and it is again at rest. It is thus a quality, a state, a condition, a phenomenon of body, but incapable of independent, self-subsisting existence. One but deludes oneself in thinking that one has any conception of motion, except as a somewhat moving. It is, therefore, incapacitated from taking rank in thought as an eternal, self-subsisting reality in its own right.

Of Mass, we have seen all along in speaking of it, that it finds its essence in pure resistance, *i.e.* inertia. But resistance is a negation, and from this fact alone, if from no other, Mass is also disqualified for an independent, self-subsisting existence. And yet it is the one, and the only one, ultimate property or potency of matter. Helmholtz's and Thompson's 'perfect fluid' must have it, and so also Le Sage's 'ultra-mundane corpuscles'; but to *have it*, is a very different thing from to *be it*.

Mass and motion thus having no right to an independent, self-subsisting existence, it follows that mechanics, and with it the whole round of science, is compelled to find its ultimate ground in the psychical side of personality. This is in no wise to discredit science, or its methods; and it is only just to say that the leaders of science clearly see, and freely admit, the position. As Professor Huxley points out, physical science must recognize its obligations to metaphysic, and metaphysic, on the other hand, has no right — indeed has no power to speak intelligently except in the fullest recognition of the province and functions of physical research. These two phases of reality stand towards each other much as the mechanical and psychical factors stand toward each other in personality, — neither being able to dispense with the other, though the psychical mode claims, and must always be granted to have, the dominance. Perhaps the radical mistake which lies at the bottom of the long-standing antagonism between these two phases of thought, is the failure to recognize the fact, that neither the mechanical nor the psychical mode of existence can claim any exclusive title to reality, — that reality as Thing, and reality as Thought, are both equally real, neither of them having this title in their own right, but both actual, as having their ground in the One, Infinite, and eternally self-subsisting Personality.

At the risk of repeating, let us look at this a little further. I am not conscious of my bodily existence apart from and independent of my psychical activities, and I am not conscious of the psychical side as in any wise independent of my body. I can easily emphasize the one or the other, and so, for the moment, throw out of account the neglected factor; but the moment I attempt to examine my thought, the whole self demands recognition. Thus it is, that neither of these two necessary factors can usurp self-ness or personality. They are either of them manifestations of 'me,' and by no possibility can the 'me' evacuate in favor of either. I speak of, and must speak of *my* mind, and *my* body, and so inevitably assume the personal subject as the very reality to which they both belong, and for and through which they both are real.

This being true of that which we may be safely assumed to know best — body and spirit (spirit being understood as synonymous with all that goes to make up the psychical factor of the self), and these two factors being necessary, constituent actualities of the cosmos, the presumption is not violent that what is true of a constituent element, is true, as a law, of the whole.

We can take one step further at least. No one will dispute that there are universal truths which are not material, or perhaps better stated, that no truth can be material. No one can think that the principles of mathematics and logic are what is commonly called substance. Further than this, no one can deny that the principle of vitality, carrying with it all three of the fundamental modes of personality, is exhibited throughout Nature. We do, therefore, incontestably see both the mechanical and spiritual factors everywhere in the cosmos. More than this, it is universally conceded that there must be an ultimate existent, which is the ground of both these realities. We have, therefore, in the cosmos all the conditions present, and it only remains to call this 'Ultimate Existent,' Person, to have the Macrocosm of man the Microcosm.

Again, whether the external world is what it seems to be, or whether it is at all, we cannot deny that it at least persists in *seeming*. What is this but to postulate the psychical factor of the Existent? Now, if we have to take leave of the one or the other of these factors, without doubt we shall have to hold fast to the psychical factor, and let the material go. Thus it is that we must concede the dominance of the psychical over the mechanical mode; or, if one or the other must be All, we shall have to conclude that the All is psychical.

This is a matter of so much consequence that I venture to put it once more in a little different form. We have seen, from the look we have had into the reality of 'thing,' that we could find no definite, hard, and self-existent 'stuff' out of which the world is made; and that, in the last analysis, even if it be assumed, against the trend of science, that there are essential and ultimate atoms in the ordinary sense, there must be still behind these a power of 'manufacturing'—of shaping, endowing, and moving them which presupposes meaning and intelligence,—that is, Personality.

But passing this, what ground have we in science to think that the two necessary postulates, Mass and Motion, are material? Motion we have seen to be a quality or condition, and Inertia or Mass, a reaction. Neither can subsist without a 'ground,' deeper than itself. Now, if science can find no place for the hard, inert, dead stuff, commonly supposed to compose matter, in the concept Mass, what possible reason can there be for holding it to be the ultimate 'ground' itself? And besides, any theory of external reality demands 'energy' to produce action; and the only kind of energy indisputably known to us is personal or self-energy.

Then, is there not better reason for thinking that all action and reaction which is comprised in the scientific postulate of the 'conservation of energy' finds its reality in a somewhat which we do actually know to be a necessary factor of the cosmos, than in a somewhat which we do not know, and have much reason to deny a possible existence? From any one of these several points of view, materialism, as commonly understood, is impossible; and it is not therefore surprising that no recognized leader of science commits himself to it; and there is no one, perhaps, who would not convict himself of a contrary belief if he could be subjected to a cross-examination.

It may be well to reassure the reader that neither science nor metaphysic has any power or purpose to spirit away the solid and substantial world he is accustomed to, and substitute in its stead some sort of mazy dream-world. Whatever may be the conclusions of philosophy, the world will continue to be, after all, just what we have all along known it to be — all too substantial and hard for some of us. Matter, such as the mathematical physicist discovers in 'thing,' and spirit, in the metaphysical sense, do not seem to differ; and if that sort of matter is thought-stuff (and must it not be from a materialistic point of view?), it must have Personality in or behind it, and is thus just what most people have meant all along by spirit. But the difference is world-wide, if the wrong end is fixed upon as the *prius*, and Personality made but an attribute or phenomenon of either matter or spirit. So long as it is recognized that Personality is the primordial reality, to which the real in any form or manifestation whatever must look for its 'being' or meaning, no harm would result from a change of names, however impossible of accomplishment. The thought-world and the material world, as we know them, would remain unchanged and we should simply have the cosmos as it is. The important point is not to fall into the error of assuming Personality to be but a quality, an attribute or a product of the cosmos. That it cannot be so may be put syllogistically, thus: An attribute or phenomenon cannot be conceived of as doing, knowing, or feeling anything whatever, but Personality acts, knows, and feels; therefore Personality cannot be a phenomenon.

CHAPTER XXVIII.

SOME OF THE GREAT METAPHYSICAL SYSTEMS.

Idealism. Fichte. Lotze quoted. Schelling. Hegelianism. Hegel quoted. Objections to Absolute Idealism. Lotze's position commended. The Supreme Good.

IT is possible, and the attempt has been made in different forms, to consider all things and events to be but the subjective modes or habits of the self, — holding that 'a non-existent world' is simply mirrored before us. This is 'Subjective Idealism,' and in modern times has Johann Gottlieb Fichte for its father.

Stated in this naked way, it seems absurd enough; but unsatisfactory as it is, one who is at the pains to understand the good and great Fichte will have small room for contempt. Lewes, an avowed empiricist, and therefore quite out of sympathy with the whole school of transcendental philosophy, says, in his "History of Philosophy": "Let us at the outset request the reader to give no heed to any of the witticisms he may hear, or which may suggest themselves to him on a hasty consideration of Fichte's opinions. That the opinions are not those of ordinary thinkers, we admit; that they are repugnant to all 'common sense,' we must also admit; that they are false, we believe: but we also believe them to have been laborious products of an earnest mind, the consequences of admitted premises, drawn with singular audacity and subtlety, and no mere caprices of ingenious speculation, — no paradoxes of an acute but trifling mind."

It has been remarked that Fichte's system is one absolutely

refusing to be compressed with intelligibility, and he must be an uncommon man who can confidently affirm that he has fully mastered it. The whole of 'Speculative Philosophy' in modern times finds its starting-point in Descartes. Decidedly the most important epoch after Descartes is found in the Critical Philosophy of Kant; but Kant's work was in the beginning, and remained substantially just what he calls it, Critical. He did not attempt to erect his philosophy into a system. He did much to make clear the transcendental, or non-empirical factor in cognition, and to point out the conditions under which rational thinking is possible; but these conditions were established, not from the nature of the ego itself, but rather from empirical sources. He did not let go a firm belief in a real objective content in our knowledge of the external world; that is, he always held that the material world exists independently of any mode or state of the ego. But yet he opened wide the door to Idealism, and Fichte was not slow to enter. The 'thing-in-itself,' which in his philosophy supports the phenomena of all bodies, he holds to be inconceivable; and essential matter is left without anything by which it can be identified with gross matter. The step is not a long one to the denial of all objective reality; and Fichte was greatly surprised that his master should not only not receive his proposed contribution with favor, but reject it with small ceremony.

Fichte declared that Kant had prepared the way and the materials for a philosophy; they needed systematizing and co-ordinating, and this was the task he set himself in his "Theory of Science" (*Wissenschaftslehre*). It would be out of place to attempt any analysis of his method. We must content ourselves with a word as to the general result. He begins with the individual ego, but as he goes on, he develops the fact that the real basis of his system is the Absolute Ego: from the Absolute Ego (God) spring all the individual egos. God is the Infinite Energy, — the Infinite Thinker, — who becomes conscious of Himself by self-diremption into individual egos. The Individual ego knows itself by the reaction of the non-ego, which is itself but a self-limitation of the Absolute Ego. The paramount principle over all is the Will.

It is confessed on all hands that the theory of 'Subjective Idealism,' though it may not be true, is impossible of successful refutation. Whatever may be the nature of that which produces cognition in the self, we know it only through such cognition; and resist as we may, cognition is the sole witness, and we cannot get beyond it.

Lotze says: " The demonstration of the 'thorough-going subjectivity of all the elements of our cognition,'— sensations, pure intuitions, and pure notions of the understanding,—is in no respect decisive against the assumption of the existence of 'a world of things outside ourselves.' For it is clear that this 'subjectivity of cognition' must in any case be true, whether 'Things' do or do not exist. For even if 'Things' exist, still our cognition of them cannot consist in their actually finding an entrance into us, but only in their exerting an action upon us. But the products of this action, as affections of our being, can receive their form from our nature alone. And, as it is easy to persuade ourselves, even in the case 'Things' *do* actually exist, all parts of our cognition will have the very same 'subjectivity' as that from which it might be hastily concluded that 'Things' do not exist.

"The assertion that the World is the creation of his own energy in his imagination could not possibly be accomplished with complete freedom from obscurity by any one except some lone individual indulging in philosophic speculation. Since it is quite too absurd that this one person deemed the remaining spirits, too, in whose society he is conscious of living, as merely products of his own fantasy; and since rather the same kind of reality for all *spirits*, at least, must be credited, therefore the question arises: How do these individual spirits, A, B, C, D,...

come to produce, by means of their faculties of imagination, four (or, if the case require it, *n*) pictures of the world, which have as a whole the same content, but which so vary in their particular features, that the other spirits, B, C, D, . . . appear to A at definite places, and A in turn to them at another place; in brief, that all appear to each other in such manner as to make it possible for one to seek for and to meet with the others, for the sake of a mutual action in this non-existent phantom world?"

In the hands of Schelling, Fichte's Idealism undergoes certain transformations. The 'object' with Fichte had reality, it is true, but it depended entirely upon the Absolute Will; the non-ego was the product of the ego, and so had no content in itself. Schelling identifies subject and object, and gives us what is called Objective Idealism. "Nature is spirit visible; spirit is invisible nature." Schwegler epitomizes the earlier views of Schelling as follows: "The first origin of the conception of matter springs from nature and the intuition of the human mind. The mind is the union of an unlimited and limiting energy. If there were no limit to the mind, consciousness would be just as impossible as if the mind were totally and absolutely limited. Feeling, perception, and knowledge are only conceivable as the energy which strives for the unlimited becomes limited through its opposite, and as this latter becomes itself freed from its limitations. The actual mind or heart consists only in the antagonism of these two energies, and hence only in their ever approximate or relative unity. Just so it is in nature. Matter as such is not the first, for the forces of which it is the unity are before it. Matter is only to be apprehended as the ever-becoming product of attraction and repulsion; it is not, therefore, a mere inert grossness, as we are apt to represent it, but these forces are its original. But force in the material is like something immaterial. Force in nature is that which we may compare to mind. Since now the mind or heart exhibits precisely the same conflict, as matter, of opposite forces, we must unite the two in a higher identity. But the organ of the mind for apprehending nature is the intuition which takes, as object of the external sense, the space which has been filled and limited by the attracting and repelling forces. Thus Schelling was led to the conclusion that *the same absolute* appears in nature as in mind, and that the harmony of these is something more than a thought in reference to them. . . . The world is the actual unity of a positive and a negative principle, 'and these two conflicting forces taken together or separated in their conflict, lead to the idea of an organizing principle which makes the world a system, in other worlds, to the idea of a world-soul.'"

Without attempting to follow Schelling in the development of his subtle and elaborate system, we have only to say that he seems to have lost himself at last in the mazes of mysticism. His earlier position, while not free from difficulties (can any system be?), was stronger, and doubtless nearer the truth. His latest enunciations, after a long period of silence, were confused and strongly pantheistic.

We are quite conscious of the fact that pantheism is a charge easily made, and that any system is in some sort open to it. There is a right pantheism, and a wrong. So long as Personality is clear and conspicuous, no system can be offensively pantheistic. It becomes so when this primordial fact is lost or confused, — when the world comes to be what it appeared to Professor Teufelsdröckh at that great crisis in his life, when he declares: "To me the Universe was void of all Life, of Purpose, of Volition, even of Hostility: it was one huge, dead, immeasurable Steam-Engine, rolling on, in its dead indifference, to grind me limb from limb. O the vast, gloomy, solitary Golgotha, and Mill of Death! Why was the Living banished thither, companionless, conscious?"

The system of 'Absolute Idealism' (Hegel) has perhaps

made most stir in the world. Schelling and Hegel began their work together, or more accurately, Schelling, who had some years the start, extended a hand to the younger philosopher, and accepted him as his coadjutor and peer. They parted company after a time, the younger soaring on to a brilliant height, the elder suffering a partial eclipse.

The system of Hegel is simply astounding in its logical astuteness and obscurity. Nobody presumes to question, or understand, its philosophic sweep. It has come to be an accepted pleasantry that it is unanswerable, because nobody dare say he fully comprehends it. The master himself put an estoppel on his disciples when he declared: "One man has understood me, and he has not." How then shall any who is not a disciple dare presume?

But, understood or not, Hegelianism has deservedly exerted a powerful influence upon the thought of the world. It is now no longer a school in Germany; in England and America, efforts partially successful have rehabilitated it in a manner; but if we may rely upon such an ardent defender as Dr. William Wallace, "few if any profess to accept the system in its integrity."

The differences between Schelling and Hegel are serious enough, from a philosophical point of view, but can only be appreciated by those who have already gained a considerable insight into the systems of the two philosophers, while they both owe their groundwork, and much of their superstructure, to Fichte.

Hegel starts with 'Being,' which in its want of content is utter emptiness or *nothing*. These two concepts are, at the same time, absolutely identical and absolutely contradictory — either losing itself in the other. But perhaps the reader will like a plunge of half a minute into the philosopher's own phraseology. He says, in the "Logic": "If we enunciate Being as the predicate of the Absolute, we get the first definition of the latter. The Absolute is Being. So far as thought goes, this is the initial definition, the most abstract and sterile. . . . But this mere Being, as it is mere Abstraction, is therefore absolutely negative; which, in a similarly immediate aspect is just what may be said of *nothing*. Hence we derive the second definition of the Absolute; the Absolute is the naught. . . . Nothing, which is thus immediate and identical with itself, is conversely the same as Being is. The truth of Being and of Nothing is accordingly the unity of the two; and this unity is *Becoming*."

The philosopher fully appreciated the opening he gave for ridicule. He says: "The proposition that Being is the same as Nothing seems so paradoxical to the imagination or understanding that it is perhaps taken as a joke. . . . It is as correct, however, to say that Being and Nothing are altogether different, as to assert their unity; the one is not what the other is."

"In Becoming, the Being which is one with Nothing, and the Nothing which is one with Being, are only vanishing factors; they are, and they are not. Thus by its inherent contradiction Becoming collapses; or is precipitated into the unity, in which the two elements are entirely lost to view. This result is accordingly *Being determinate*, or definite. . . . To Being, therefore, in this stage is attached a determinateness (a certain cognizability) which, as it is immediate and said to *be*, is *Quality*. And as reflected into itself in being so determined, the determinate Being is *Somewhat*, in being there and then. . . . Quality, as determinateness which *is*, as contrasted with the *Negation* which is involved in it, but distinct from it, is *Reality*. Negation, which is no longer an abstract nothing, but Somewhat which is-there-and-then, becomes a mere form to Being — it is Being other than some-Being. This Other-Being, though a determination of Quality itself, is in the first instance distinct from it. Quality is *Being-for-Another* — one width, as it were, of *Determinate Being,* or of Somewhat. The Being of Quality, as such, contrasted with this reference connecting it with another, is Being-by-Self."

This will doubtless be enough for most readers. It is not jargon, as some may be inclined to think, but every sentence above is full of solid truth. Its form, however, is so uncouth, or at least, so out of the run of common phraseology, that the meaning is not obvious at first glance.

With Fichte the ego devours the non-ego — all is subject; with Schelling the ego and the non-ego both subsist, and their identity is indifference; but with Hegel the *relation* between the two forms the basis of all truth. The discovery of relation is the province of thought, and so he resolves the Universe into *Thought,* and his Metaphysic is Logic. Everything is rational, and everything rational is actual. Schwegler epitomizes his position as follows : " The ' idea ' is the highest logical definition of the Absolute. The immediate existence of the idea we call *life,* or the process of life. Everything living is self-end, imminent-end. The ' idea ' posited in its difference as a relation of objective and subjective is the *true* and *good.* The true is the objective rationality subjectively posited; the good is the subjective rationality carried into objectivity. Both conceptions together constitute the *Absolute idea,* which *is* just as truly as it *should* be, *i.e.* the good is just so truly actualized as the true is living and self-realizing. The absolute and full idea *is in space,* because it discharges itself from itself as its reflection; this its being in space is Nature."

A true Hegelian is apt to look with a certain degree of intellectual compassion upon any who, pretending to think at all, do not embrace the Hegelian system in its fulness. Whatever is true, no matter if it be as old as Parmenides and Heraclitus (to whom Hegel freely acknowledges his obligations), they seem to insist upon seizing to the behoof of their Master.

There was philosophy before Hegel, great as he undoubtedly was, and his system does not close the circle of thought.

Serious objections are urged against the system. In the first place, with regard to method: he, beyond all the Idealists, starts with the most 'abstract and sterile' concept — the Absolute which he defines as Being or Naught — and from this, without break, he professes to deduce the World; which must, of course, include 'Thought' in which all things find reality. Does he do it? Does he not rather drop down into the known world at every point to get material with which to build his Logical Palace?

With regard to his famous 'Dialectic,' the instrument he uses in constructing his giddy heights, he is clearly entitled to a caveat in modern times. It consists, in a word, of the play between Being and Naught. Before a thing can be known, it must be 'othered,' or pass out of itself into negation, and from it return as posited or known: thus contradiction is absolutely necessary to the knowledge of anything.

In the second place, he uses 'thought' in a non-natural and misleading sense — else Thought and Being cannot be identical. Thinking is an activity; it is a mode, and never can be 'thing.' It is essentially dynamic, and presupposes a subject. Thought does not think. The thinker only thinks, and the content of thought is like and unlike — relation. The claim is made, and it may be quite just, that in the Hegelian system "thought regarded as the basis of all existence is not consciousness with its distinction of ego and non-ego"; that "it is rather the stuff of which both mind and nature are made, neither extended as in the natural world, nor self-centred as in the mind"; but if this be true, it is unfair to call such an entity thought, since it differs as wide as the poles from the notion the world has, and is sure to retain, of thought and thinking.

Another, and the most serious objection, is that it mutilates Personality. Thought, in any way it may be construed, is not

Person; the self is a thinker, but this activity is but one mode of the self: the self is also energy or will, which is distinct from thought; and it is Feeling, which cannot be confounded with either: and no system can embrace the whole truth which leaves out of account, or which only smuggles in, Personality.

Again, the system of Hegel attempts to do what we regard as unphilosophical and impossible. Instead of confining himself to the actual world as discoverable in the Universe as it is, he attempts to transcend the domain of all actual knowledge, and establish laws which must have governed the development of the cosmos, or which the Creator must have followed in his own creation; and after he has his Logic-world, the Creator has no proper part or function in it. He is at last an abstraction, — the naught.

Goethe says, somewhere, Man is not born to solve the mysteries of existence, but he must, nevertheless, attempt it, in order that he may learn to keep within the limits of the knowable. The problem of how to construct a World is not proposed for our solution, and is inconsequent and presumptuous. We must recognize the fact that the world is a revelation to us, as we are a revelation to ourselves. We must accept what, by the conditions of our being, is forced upon us; and although we may take upon our lips great swelling words of doubt and denial, march against high heaven, and assault it by logic or by hate, we but strut and vapor in an idle show, and return at last to that which is given us in the world as it appears to us through the self.

Philosophically, we take our stand humbly by the side of the gentle, full-souled Lotze. He says: "If things exist and events happen simply in order that the formal relations of Identity and Opposition, Unity and Multiplicity, Indifference and Polarity, of Universal, Particular, and Singular, etc., may be actualized in the most manifold manner possible, and set forth in Phenomena; then, of course, the essence of 'Things' is so pitiful and insignificant that our *thinking* succeeds perfectly well in adequately comprehending it.

"The teaching of Fichte had been different. The problem of Spirit, he held, does not lie in the cognition of a blind Being (the conception of which appeared to him as impossible as it appears to us), but in action. The aforesaid world *is* not, but appears to us in order to serve as material for our duty, as inducement or object of our action."

For the 'action' of Fichte, Lotze proposes to substitute the Good — for which 'action' is simply the indispensable form of actualization; in which supreme concept is included the 'Beautiful,' the 'True,' and all blessedness — uniting into one complex whole all that has *Value*. "And now," he says, "we affirm: Genuine Reality in the World (to wit, in the sense that all else is, in relation to It, subordinate, deduced, mere semblance or means to an end) consists alone in this Highest Good personal, which is at the same time the Highest Good Thing. But since all the *Value* of what is valuable has existence only in the Spirit that enjoys it, therefore, all apparent actuality is only a system of contrivances, by means of which this determinate world of phenomena, as well as these determinate Metaphysical habitudes for considering the world of phenomena, are called forth, in order that the aforesaid Highest Good may become for the Spirit an object of enjoyment in all the multiplicity of forms possible to it.

"The objectivity of our cognition consists, therefore, in this, that it is not a meaningless play of mere seeming; but it brings before us a World whose coherency is ordered in pursuance of the injunction of the Sole Reality of the World, to wit, of the Good. Our cognition thus possesses more of truth than if it copied exactly a world of objects which has no *value* in itself. Although it does not comprehend in what manner all that is phenomenon is presented to its view, still it understands what is the *meaning* of it all; and is like to a spectator who comprehends the esthetic significance of that which takes place on the stage of a theatre, and would gain nothing essential if he were to see besides the machinery by means of which the changes are effected on the stage."

It is impossible that there shall not constantly arise in our minds a demand to know more of the Good (only another, if another, name for God) than can ever meet with adequate response. We cannot but ask, in some inarticulate way, even when we know that the answer cannot be given except in terms which degrade the Ultimate, in the very act, from its awe-inspiring reserve, Where is it? How is it? What is it? And it is doubtless because the answers given only too readily by certain classes of Theologians — who, somehow the world seems to think, ought to know — have been so definite and exact in 'words without knowledge,' and which men in our day have come to recognize as narrow and inadequate, that so many deprecate the notion of personal existence in the Infinite, and attempt to find more satisfying forms of such existence in ideas of an Eternal World-Order, an Infinite Substance, or a Self-developing Idea.

We must allow ourselves a few words from the "Microcosmos" touching at least one of these proposed substitutes: "What noble motives and what moral earnestness may lead to the dissolving of the Divine Being in that of a Moral World-Order, as contrasted with crude anthropomorphism, must be still fresh in men's remembrance. And yet Fichte was not right when, with inspired words, he opposed his own sublime conception to the common, narrow-minded idea of a Personal God; because he sought that which was most sublime, he thought that he had found it in the conception which he reached; if he had followed out to the end the path which he took, he would have recognized that by it that which he sought could not be reached. The question, *How* is it that a World-Order can be conceived as the Supreme Principle? cannot be put off by appealing to the fact that we cannot demand a history of the origin of the Principle itself; he who, regarding Personality as an impossible conception of the Godhead, prefers some other to it, will at least have to show that the one which he brings forward is not contradictory; for nothing will be gained by substituting for an impossibility some other assumption of which the possibility is not proved. Now the fact is, that the one sufficient reason which will always forbid that some World-Order should be put in the place of God, is to be found in the simple fact that no order is separable from the ordered material in which it is realized, still less can precede such material as a conditioning or creative force; the order must ever be a relation of something which exists, after or during its existence. Hence, if it is nothing but *Order*, as its name says, it is never *that which orders*, which is what we seek, and which the ordinary notion of God (however inadequate in other respects) determined rightly at any rate in this, that it regarded it as a Real being, not as a relation."

We cannot pursue the subject further. All efforts to empty the Universe of a Personal Author and Sustainer must face the rebuke of the Hebrew Prophet, "Shall the work say of Him that made it, He made me not? or shall the thing framed say of Him that framed it, He had no understanding?"

CHAPTER XXIX.

ETHICAL.

Self-activity necessary to Morality. No Liberty in Sensibility or Cognition, as such. Choice. Motives. The 'Good.' Obligation. Man held to be Omniscient. No Obligation in Selfness. Altruism. How the Will of the Supreme Good is known. The ' Categorical Imperative.'

IT is idle to dispute that there can be no moral good if there be no liberty of action. As a universal fact of consciousness, no man is held to be the author of a deed if it be known that he was absolutely compelled at all points to do the deed. Indeed, we do not even say, in such case, that *he* did it. For example, suppose one were taken by force and placed upon one's knees before an idol; it would be sheer perversion of words and sense to say that that man knelt to the idol; it would be impossible to think of him as in any wise responsible for the action. And this must be true in all cases where the possibility of self-activity is absolutely cut off.

Now there is no question here as to whether we are bound ultimately, as some hold to be the case, by a rigidly predetermined order, or not. All that we say, for the purposes of this argument, is that if we are so bound, in every particular, we are not responsible, and the actions attributed to us are in no sense ours.

Now of the three fundamental modes in which personality manifests itself, — sensibility, cognition, and volition, — two, as such, are absolutely bound. Sensibility is, if we may so say, the motive or propelling power; cognition, the knowing or intellectual power; while it is reserved for the will to give direction and control our actions.

That the sensibilities are compelled to receive whatever is impressed upon them by stimuli, without the possibility of the sensibilities themselves varying their reactions in response to such stimuli, is easily seen. In a given state of my visual organs, and with my eyes fixed upon a page, can they create or drive away the characters which I see? If a sharp instrument be thrust into my flesh, is the pain of my making? can I bid it begin or cease? A rose is brought within the radius of my olfactories; am I at liberty to perceive its perfume or not? So long as I am conscious, I am compelled to suffer just such pain, or perceive just such agreeable affections, at that moment as my nervous organism in its then state is competent to reveal.

The case is equally obvious with respect to the intelligence. It, too, is fast bound. By the understanding I am made to know the meaning of the words on the printed page; that the instrument is sharp, and has pierced my flesh; that the rose has an agreeable perfume. Can it, in its assumed state, do less or more? Can it tell me that the word 'rose' is composed of more or less than four letters, or that it is not the name of a flower? Can it tell me that the sharp instrument is a perfume, or the perfume a bodkin? It must render to consciousness, at any particular moment, just such report as the degree of attention and its then power is able; and that as absolutely as an instantaneous photograph must respond to the conditions of light and shade at the instant of exposure. It tells me that what I read is interesting or dull, melancholy or humorous, true or false, as it is competent. True it is that different intelligences, or the same intelligence at different times, pronounce divers judgments upon the same thing. It is often deceived: sees what is false as true, and the true as false; it may be confused or in doubt; but it has not two voices at the same instant. For the nonce it is what it is, and has no power in itself to change the result. And thus thought, as such, has in it no possible element of liberty.

It is, therefore, vain to look for the ground of morality in either thought or feeling as such; and any system of ethics founded upon Pleasure, or Happiness, or the 'Fitness of Things,' or Utility, or anything whatever which reaches no higher than thought and feeling, is founded upon a misconception of the psychical conditions of personality.

Where, then, shall we look? We have only the Will left; and if moral accountability exist at all, it must depend upon this master-mode of the self.

We have seen abundantly already that sensation and thought cannot be numerically separated from the will, nor the will from either of them. It is not enough that the eyes be open, and the page exposed, to read. There must be attention, in one degree or another. But what is it to attend? It is to energize or exert power. There is energy of some sort in all thinking and feeling, even in the illustrations used above, in which the dominating elements, sensation and understanding, were emphasized; there was also present, of course, some degree of volition. When I withdrew from the sharp instrument it was not sensibility that acted, even though the movement was automatic; nor when I discovered, as was possibly the case, that there was a mischievous urchin behind the pin, it was not the intellect that administered chastisement. It was the self, comprehending the state of the case through sensation and understanding, which put forth the volitional energy.

It is not necessary to go over again the question of this free activity. It is simply a fact of consciousness forced upon us, affirmed in consciousness in the very act of denial, and just as certain as that there is anything to deny.

One point, however, we must consider for a moment, and that is, the action of motives upon the will. Motives there are, though it is quite certain that they are not always discoverable in consciousness, as is shown by the constant declaration that one does not know why one did or said this or the other thing.

In them, as such, we have seen that there is no freedom, just as there is none in all that follows the purposive epoch — the actualization of an act of will. But the question is this: Can any one honestly declare that the motive which governed his action in any case, he not being in a hypnotic or other abnormal state, was so strong and so definite in consciousness that he could not have disregarded it? It may be answered that even if one felt that one could have set it aside, one would have had a motive for doing even that. This is clearly to shift the point. The question is not whether one *is* so bound that no action is really his own, but whether he *knows* and *feels* himself so bound that the actions which are attributed to him are not his, but are the result of an irresistible compulsion consciously due to environment. There can be but one answer; and so whether, as known to a hyper-human being, we are actually free to will or not, *we* feel — and in this sense *know* — that we are; a knowledge just as certain as that there is an external world. It is a fundamental fact, revealed to us in the only way anything can be known to us, and which admits of no practical dispute, whatever speculation may suggest.

But all free activities are not, therefore, moral. It is necessary, then, that we should arrive at some definition of the 'good.' From what has gone before it will appear at once that we must look for its ultimate ground in the one Infinite Personality, — the good for us being that which is in harmony with his laws, and the evil, that which is in conflict therewith.

But there are two obvious senses in which we may use the word 'good.' It may mean anything which promotes the growth or development of 'thing'; as the building up of tissue in the body, in the animal and vegetable worlds; or causes any economic or esthetic change in the material world. We speak of good health, good food, good fortune, and in general, of anything which has value, that is, ministers to the happiness and general well-being of man. These all fall under one or other of two heads, — the pleasurable or the useful. Attempts have been made to find the end of all human effort in either of these; giving rise to the systems of Hedonism, with Hobbes as its most notable exponent in modern times, and Utilitarianism, with Bentham as its chief exponent. They both, together with a number of other phases held by eminent writers, may be classed under one general head — selfism.

The other sense of 'good' is synonymous with virtue. This is moral good, and finds its essence, not in the end of the action, but in the act itself.

But as there must always, in any action, be an end towards which the effort tends, in virtue, the end is never self, but the not-self. This is now commonly called Altruism. The essential difference between these two kinds of good lies in this: selfism looks inward; Altruism looks outward.

In selfness there can be no obligation; for no law can emanate from the self, which the self cannot at any time set aside. We do indeed speak of "being true to oneself"; but in any such case we shall find, upon analysis, that we mean to hold the self true to some rule or principle which has its reality over and above self. In any system which has one's own betterment or gain, or satisfaction in any form, for its motive, we must recognize an inflow towards the centre and source of action. It is acquisitive in its nature, and the only law which governs or moves the actor can look no higher for its source and authority than the individual personality itself; and thus cannot be binding in the slightest degree one moment longer than held to by the self. For such an action the supreme power is in the self, and it must be able to loose as well as bind.

In a selfless, or altruistic volition, the flow or movement is from within, outward. There is a recognition of a law laid upon the self from a paramount authority, and so obligation emerges. There is a recognition of a 'Law-Giver,' who is paramount to the self, and 'duty,' 'ought,' 'accountability,' gain their meaning. This source of law with the child begins with parents and masters, and rises through the stages of all human law, express and implied, in society and government.

But one soon finds that there is a source of law which is the 'ground' of all human laws, — a source above and beyond the temporal — proceeding from the Author of all rule and authority — the Infinite Law-Giver. His laws come to us in a twofold order: one the law of mechanism; the other the law of personality. We call the one Nature, the other Spirit.

Now Nature, or Mechanical law, bears in her hands her rewards and punishments visible and open. Her voice is, Do this and you shall surely have your pay in current coin: or, If you do not this, behold the rod! It is a system of open and avowed rewards and punishments. One who follows the bent of a desire does it because he expects the gratification which attends it. One who takes bodily exercise does it because he expects increased health and strength: one who, relying upon nature's law of compensations, gives up leisure for toil, expects return in money, in skill, in learning, in power of one or other of its thousand sorts. He does not, it is true, always get what he expects. That would be too much, — that would mean the entire satisfaction and saturation of his nature; he would not, and could not, look further. But he does get just what nature promises, and what he knows she promises, if he be at the pains honestly to inform himself. He gets satisfaction up to a certain limit, more or less sharply defined; but he knows that there is such a limit, and he knows when he is passing it; but, for the sake of some remaining sparkle, he is willing, too often, to drink the dregs which he clearly sees.

This pleasurable or happiness stage, in the order of nature, does not wear the rigorous aspect of law. There is small token of compulsion, — her aspect is one of smiles and enticements. This, because she must play the nurse to man's higher life; and, in order to lead him forward to a proper knowledge of his self-developing and self-governing functions, there must be more caresses than cuffs. But, if she is felicitous and gracious, she is at the same time firm and inexorable. She is sure to show the danger line; and if it be not heeded, we must take the consequences. Nor does the plea of ignorance avail. If one, under the firm conviction that one is sweetening his tea with sugar, should use arsenic instead, the result would not be less fatal because the result was not foreseen. If one should walk off a piazza in the dark, one's confidence that there was no danger would not save a broken limb. So, too, if an engineer build a bridge out of bad materials, thinking them good, or upon wrong calculations of the strains, calamity would result as surely as if he had intended to produce the disaster. Thus it is that nature holds every man to the same account as if he were omniscient. She gives us the power to inquire into, and find out, her ways, and abundant warning that she makes no exception in her mechanical order, and that we act — must act at our peril.

In all this phase of nature's order, we are but accepting her gifts, — using them rightly to our profit, or wrongly to our hurt. Clearly there can be no right or wrong, in a moral sense, since we are in the attitude of beneficiaries, — accepting her bounty to our own gain, or abusing it to our loss: and this embraces the whole sphere of our sensuous, intellectual, and esthetic life.

But this realm of 'things' is not all of life. This is but the acquisitive — the inflowing phase of man's existence, in which self is consciously the object and recipient, in mind and heart, of the good gifts of the One Infinite Source of All Good. These gifts are good, in a right sense, only as they are regarded in the light of this Infinite Activity, — good because they are given, — because they issue forth from the Absolute Personality. They are good *to* us, because they meet the uses and desires of our nature; but not *in* us, because they are but the accidents and occasions of self-activity and self-development.

The radical fault in all systems of morality founded in self-ism is, that they stop in, and cannot by any possibility rise above good, in this reflected or borrowed sense. They rest in good *to* man from God, and cannot logically, in any possible way, find room for good *in* man, for the love of the good — which is only another way of saying — for God's sake.

The second and higher aspect of Law, which we have called the Law of the Spirit — for and through which the lower phase is entitled to reality — is the exact contradictory of this Law of Nature. In it there is an outflow, in obedience to a recognized obligation laid upon self. It is a free activity, rendered possible by the fact of man's power of self-determination. Though infinitely less in degree, it is of the same nature as that natural good, which we have been considering, regarded from its preternatural or divine side. That looked from God toward man; this looks from man toward God. God gives to man; man, by the power given him, gives to God. God, as absolutely free and of infinite power, is All Good. Man, as bound about on all sides, is free only in the purposive epoch of the will. He is in this one respect, in the 'image' of his Creator; so that as God is the source and ground of All Good, man by this power of self-activity — itself a gift to him — is competent to be the source of *some* good. And just as all Natural Good — the World-Gift, is the creative out-go from the Infinite, the All-Giver, so the limited, the little good man is competent to, is the purposive outflow from self. This is morality; — this is self-lessness.

With this principle well in view, we can have no great difficulty to determine, at least theoretically, the moral quality of any action. Whatever is for the betterment of another is moral; whatever is for the sake of self, without prejudice to another, is morally indifferent, and will be profitable or harm-

ETHICAL.

ful according to whether the judgment is sound or faulty, assuming it to be honestly followed. If not honestly followed, it will be, in so far, hurtful. That which is purposely hurtful to another is wrong, — is evil.

With regard to self-less or moral action. As we have seen, it must be altruistic, that is to say it must have an object which is not self, and the purpose must be the betterment of such object. But the moral quality does not lie in the object, nor in the actualization which follows the purposive epoch. The intended good 'thing' may utterly fail of the purpose by accomplishing nothing, or even by working a positive hurt to the object. For example, one with a benevolent purpose may buy bread to feed the poor. It may never reach them, or reaching them, may prove to be unwholesome, or even poisoned, and so work destruction. Such results do not in the least affect the virtue of the volitional factor, and so the action itself. The moral quality, thus, lies neither in the material of the action, nor in the result, but in the purpose. The fact is, that both the material and the result are mechanical. It is only the will in the purposive epoch that is free. The motions of hand or tongue are neither of them free; for once the purposive energy of the will is put forth, the motor-nerves and the muscular response are in no essential point different from the blind action of a machine. The tongue or hand may not move at all, or may move in a different way from that intended, entirely dependent upon the degree of excellence in the bodily organism.

In the same way, the result of the movement, in its immediate aspect, is movement only. Whatever psychical effects may follow will be mediate or secondary, depending directly upon the use made by other personalities of such movement as a stimulus.

Now the object towards which this free-purposive energy is exerted in a moral effort must be, either other individual personalities, or the One Infinite Person. With a child in the early stage of moral development, it will be mother, nurse, playmates, or even pets, towards whom this energy is exerted; but after a certain stage of moral discernment is reached, largely dependent upon education, it will be discovered that there is a Power above and beyond all these which is the source of all good; — and from this time forth, individuals and the All-Father can never be disassociated in a moral action. No good 'thing' can be purposed for the betterment of any, which does not accord with the more general purpose to will in harmony with the Infinite Will: and wherever there is a temptation towards such divorcement, the Infinite must take precedence, or the effort ceases to be virtuous.

But now the question arises, How can we know the will of the Infinite Good? The answer is, We know it just as we know anything else which is disclosed to us — that is through the understanding. By virtue of the primordial order written large upon the nature of the self, and of the capacity given in accord with it, we must understand or co-ordinate the phenomena of the world about us; and as we gather knowledge of 'things,' little by little, so we must begin to construe actions, as well in the realm of the 'good' as in the realm of 'things.' In the latter sphere in which we distinguish the mere useful or pleasing, and in which there is no necessary moral quality, the understanding is not faultless, but varies from time to time, — pronouncing that useful, pleasing, or true, which, upon more reflection or in better light, is seen to be hurtful, impleasing, or false. In like way, the understanding has no claim to infallibility in its judgments in regard to the morally good. And thus it is that we see a wide difference, in many points, as to what is held to be good among people of the same general environment, as well as in those widely separated in space and time. But as there is a large area of general agreement, with a narrower circle of absolute and necessary agreement in the judgments of the understanding in the sphere of sensuous and intellectual truth, so there is a large area of general moral truth, with a correspondingly narrow inner circle of absolute moral convictions; and the agreement in the universal consciousness of men is not measurably larger in the one case than in the other.

For example, in the sensuous and intellectual domain, people differ as to the agreeability of the flavor or perfume of fruits; as to this or the other theory of economics, this or the other explanation of geological, astronomical, or chemical phenomena, with a large area of agreement. They agree entirely upon the fundamental principles of mathematics.

In the domain of moral truth, they differ as to the morality of horse-racing, cock-fighting, card-playing, the use of intoxicating liquors, and many like points. They agree as to theft, cruelty, injustice, selfishness, and all that can be called essential. There is no more doubt of a general consentient voice in moral than in intellectual truth.

That a moral action has a different character from all other actions is chiefly attested by the fact that consciousness compels us to recognize an obligation to put forth this purposive energy in the case of moral action, and in no other. The 'thou shalt,' 'thou shalt not,' are laid upon the self. This is the 'Categorical Imperative' of Kant. The question as to whether it is a fact of personality, or not, has to be carried to the tribunal of every man's consciousness. That the answer can be but one way, is attested by the languages of all peoples, by the existence of human law, and by every man's actions. There is no one who does not use the word 'ought' in some form of expression or other, and who does not hold his fellow-man to an accountability. It does not persuade nor counsel, but commands, and we recognize its authority.

It is a most significant fact, that this sense of obligation attaches to no other class of actions than the moral. When an act lacks the before-mentioned characteristics of morality, it may be done or left undone, and the word 'ought,' in its right sense, does not fit the case. No obligation is felt to take one, rather than another, of two walks, when the object is mere recreation; or to eat of a particular dish at table, to sit, stand, run, or do any other act, when the relief of or injury to another is not an issue. But any or all of these may become moral by making them touch, in any serious way, the well-being of others; in such case obligation immediately attaches.

And now the further question presents itself: Why should the class of actions which we call moral, enjoy a higher prerogative in consciousness than those which are merely prudent or expedient? The simplest answer to this question is, Because they do. It is quite analogous to the question as to why the whole is seen to be greater than the part. That and this are presuppositions of Self-nature — intuitive and immutable laws of our being. From the relative point of view, they are entitled to this pre-eminence because they touch the supreme destiny of man as an individual and a race.

CHAPTER XXX.

THE NATURE AND FUNCTIONS OF CONSCIENCE.

The admonitions of the moral monitor. Conscience discovers itself only upon change of moral purpose. Analogy between the functions of conscience and inertia. Analysis. An illustration. Moral momentum.

WE have not yet exhausted the facts of consciousness with regard to moral action. Not only does obligation discover itself to us in all acts possessing moral quality, but an admonition or penalty follows the failure to respond to an obligation laid upon us. When the understanding makes it plain to us, that a particular action which lies before us ought to be done, and we do it not — or, that it ought not to be done, and we do it, we feel a dull, heavy distress about the heart, which we can neither avert nor control. It is undoubtedly a reflex action, and is unmistakable in character. It attends no other sort of actions than those in which we are made conscious of moral obligation, and only follows a wilful disobedience of our sense of right. In this it is altogether unique. We may run into danger, make a pitiful blunder in judgment, or stake our fortunes upon a desperate hazard, and all with the worst results; but no one will assert that conscience obtrudes itself upon us under any of these circumstances. There are sensations, very distinctly discoverable on all these occasions, — confusion, terror, shame, sinking-of-heart, but none of them can be in the least mistaken for that peculiar sensation which we call 'qualms of conscience.'

This moral monitor is not, then, always discoverable: it is indeed never active where mere intellectual or esthetic questions are in issue. But is it always discoverable where there is moral activity? We think not. A right-minded man, following uniformly what he understands to be right, though constantly passing upon such questions, will rarely feel the presence of this admonitory sensation; never, if there be no departure, nor purpose of departure, from his conception of right — a state of case which could never be, except in one absolutely perfect. But where the right course is plainly seen, and the obligation to follow it clearly confessed in consciousness, if one wills not to yield to the obligation, one is sure to feel the dull thud of conscience. If one change his course from open, or intended disobedience, this action of the will is followed by a corresponding lightness of spirit, which is called the approval of conscience. Thus conscience discovers itself only upon the change, or purpose of change, from right to wrong action, when it will be deprecatory; or from wrong to right, when it will be commendatory. Its field of action, therefore, is entirely analogous to that in which causation finds opportunity — in change.

Conscience is not an illuminating, nor judging power, in any sense. It is itself perfectly blind, and always reactionary or negative. Hence we cannot say with Reid, and perhaps the majority of writers on the subject, that conscience is the 'candle of the Lord set up within us to guide our steps.' It is rather a hand-rail to keep us in the right way. It does not point out the way — that is the office of the understanding; but when we know the way, or are thoroughly assured that we do, it does not fail to protest against any departure from it. It varies in intensity with different people, without doubt, and at different times, in the same person, according to the suddenness and gravity of the change in conduct.

Now there is a striking analogy between this conserver of the moral world, and inertia, the conserver of the physical world. All bodies, by virtue of inertia, let it be remembered, resist change, either with respect to rest or motion; that is, a body at rest resists all effort to move it, and once in motion resists all effort to deflect it from its rectilinear course, or change its velocity. The importance of inertia in the mechanical world cannot be exaggerated. By virtue of it the woodman's axe is enabled to do its office; while it maintains the movement and stability of the celestial world. What it is, we do not know. It is essentially negative in character, — never acting unless first acted upon. It is the conserver of the mechanical universe. Now this is just the office which conscience performs on the psychical side of personality.

Let us examine this somewhat in detail. Inertia is purely reactionary, — does not exert energy, but resists simply. In like way, conscience does not manifest itself unless there is conscious deviation, or purpose of deviation, from the moral path upon which the self is moving; but immediately upon the advent of a purpose of deviation, it sets up its resistance. Again, a body once settled in its new path after inertia has made all the resistance competent to it, there is no further resistance — inertia becomes again quiescent. So with conscience. When the will has acted, and conscience has been overridden, the protest ceases, and is not felt so long as there is no glancing back — no entertainment of the once felt obligation; but whenever purpose becomes tremulous, the reactionary emotion of conscience begins again to act, and the will is solicited to relinquish the path taken up against its protest, and return to that which is seen to be of obligation.

Once more, it will be remembered that the line of right action, as disclosed by the understanding, is not absolute and unvarying, but changes from time to time in the same person. That which we at one time, and under certain circumstances and teachings, thought right, we, under new influences and in better light, come to see was narrow and ill-founded; and we determine to abandon the old way for the new. Now, if conscience is blind, like inertia, and knows nothing but action, it must resist the change, even from the worse to the better. Is not this in accordance with the facts of experience? The point is a delicate one, and the facts upon which to test it are scant. The change must be marked and somewhat sudden for the reaction of the moral monitor to be clearly distinguishable; but most people probably have some experience upon the point. Let us take a case of frequent occurrence in America, — that of a youth brought up under the unquestioning conviction that it is sinful to play at cards, to dance, or to drink wine. Let it happen that, after a time, in associating with people who hold that the wrong is not in these things, *per se*, but in their abuse, his own judgment parts company with that of his father and mother. Will he not upon first indulging himself in these things, now regarded innocent, discover the admonitions of this blind monitor? Many persons certainly have felt such qualms, and if there has not been sufficient time for the moral current to settle in its new bed, every one must. So long as a thing remains merely a speculative opinion, there is no action on the part of conscience. The understanding is all the time modifying the sensibilities, but it is a slow process, and very gradual. The will must actively co-operate, not only in the prosecution of the enquiries of the understanding, but in purposive determinations to follow the new light in conduct when opportunity serves. Thus the current of one's moral nature is gradually turned; but at every point on the one hand or the other, when the change is sudden, conscience makes a stand.

This accounts for the fact that some men, persisting in evil ways, are not troubled by the stings of conscience, while others following like courses are always in a flutter. In the first case there is no thought of returning to the right way, and in the second the self is moving in a constrained path, and the will is ever on the point of giving up to the recognized obligation.

But, it may be asked, how is it that men sometimes become utterly dead to the stings of conscience, or rather that conscience ceases to sting. The explanation seems plain enough. We may distinguish two stages of this condition: first, that in which by long and unswerving persistence in disregard of the still recognized obligations of right, the moral current has become sluggish, and the, at best, faint sensation has worn itself out by familiarity. There is almost no moral momentum, and therefore no reaction developed.

The other stage is more melancholy. It is when the power of seeing what is right is almost gone. The will acts upon and modifies the understanding. It cannot by a simple mandate make it see a thing at any moment other than it actually appears, but in all matters of opinion it is easily warped by the indirect action of the will. Through interest and desire it can be blunted and blurred, just as a microscope or the eye itself can be injured — put out of focus or weakened — so that by misuse or abuse it is rendered almost useless. A thing which the judgment sees to be morally bad cannot in the same moment be thought good, but the will can throw hues of desire upon it, and after a time it will appear quite different. By a long-continued tampering with honest convictions, one may come at last to have small power — perhaps no power at all — to distinguish right from wrong. "If the light that is in thee be darkness, how great is that darkness." This is that condition which one may bring upon oneself denounced by the prophet: "Woe unto them which call evil good and good evil!" Manifestly in such a moral state the conscience is dead.

When, then, is this reaction in the heart most active? In those persons, clearly, who are most scrupulous to ascertain and do what is right. They gather a moral velocity by the accelerating energy of the will, so that the slightest deviation from the way they hold to be right causes them a far greater distress than a positive crime would to one whose moral kinetic energy is small. Thus it is that one may know oneself to be in a low state of virtue, when, upon the perpetration of a wrong, there is no positive reaction discovered on the part of conscience; and, on the other hand, no one's case is desperate who finds this reflex action still strong within him.

There is still one point which demands notice. If the analogy between the part played by conscience in the conservation of spiritual verity, and that played by inertia in the conservation of physical reality be complete, then, any increase of moral energy must be at the cost of a resistance overcome; for inertia reacts to prevent increase of velocity, just as certainly as to prevent loss. That we do find such resistance can hardly be disputed. 'The last state is worse than the first.' The loss of momentum can only be regained gradually by the expenditure of a constant force; and, that it is necessary to expend a great deal of moral energy to recover lost ground, everybody knows only too well.

Let us try to make all this a little more practical by an illustration. Take a steamship, and let her be supposed in any great current of the sea, say the Gulf Stream. Now, assuming the vessel to be entirely motionless, she would nevertheless be carried forward toward the north at the rate of four or five knots an hour. But she is fitted up with engines by means of which to take on a proper motion of her own. She is also provided with compass and chart by which she can know her course. In addition to this, she has a rudder by which, through the man at the wheel keeping his eyes upon the needle, she is held upon her course.

We recognize at once in the current which carries the ship blindly forward, the steady flow of vitality, the whole trend of subconscious activities, and the general environments of life. We are all impelled by nature, without thought or concern on our part, towards a higher evolution,—within, by the action of the heart in supplying blood to tissue, the lungs in their respiratory functions, the whole nutritive system in digestion, and the world of reflex action; without, by the reaction of place and circumstance in life, family, society, and government.

In the ship's motive power, we see a likeness to the part played by feeling when fully formed in consciousness, which is the basis of personal mobility of body and mind. In the needle we recognize the line of right action as disclosed to us by the understanding,—rarely on the absolutely true meridian, and always fluctuating in some degree; but, though passing and repassing the true line, through temporarily perturbing influences, always returning to it when the disturbing cause is removed: while the chart shows the bearing of all points with respect to the grand axial line of conscious right. In the rudder, we have the power of self-direction through the free activity of the will at the helm. In the beginning of life, we are carried forward, almost wholly by the (to us) blind forces; but after a time the masterful power of purposive control, in the light of the understanding, shows itself at the helm of right action, and from that time forth, the haven toward which we sail we determine for ourselves.

But we must not leave out of sight a most important factor in the ship's economy, one without which she could keep no course and could reach no appointed haven, but would drift hopelessly at the mercy of wind and tide. It is the reaction of the water upon her keel and sides, as well as upon the blades of her propeller. She could not move an inch if it were not for this reaction upon her propeller; nor could she have any steerage way, if it were not for the reaction of the water on keel and sides. This reaction is purely negative, unseen, and perhaps, unknown to most people; but without it there could be no certainty, no safety, and no directed motion at sea. And yet its immediate office is to resist — to resist change in direction, or change in velocity. This is inertia; and this is the corresponding office of conscience in self-destiny. It is just as absolutely necessary in moral movement, and in the conservation of moral reality, as reaction in mass is in the mechanism of the outer world, and the conservation of physical energy.

CHAPTER XXXI.

THE INFINITE PERSONALITY.

Personal good implies personality in God. The Mosaic account of the origin of evil in man. Disobedience. Obedience. A class of theologians faulted. Conflict and agreement of the Finite and the Infinite. Theology. Religion. Human aspirations. Quotation from Mrs. Browning. Conclusion.

SO long as men know that 'good' is, so long they must know that God is. Good, in its only right sense, presupposes, and is inconceivable apart from, Personality; and Personality, in its highest term, is God. The one indisputable fact of the universe for every man, we repeat for the last time, is Personality. In the moment of direst scepticism, the consciousness of doubt carries with it the further and higher consciousness of self, as a necessary and precedent fact: and unless one can arrive at such a stupendous egoism as to hold one's self to be the sole and only reality, subsisting in a sublime isolation of circumambient nothingness, one must know that there are other personalities out of and beyond one's own; and that the universe is meaningless and inconsequent, except under the postulate of an Infinite and Absolute Personality — God over all.

But we have further seen, that Personality is itself a mystical tri-unity, comprising sensibility, understanding, and will — that 'good' does not lie in sensibility, because there is no element of freedom in it; nor in the understanding, because it, too, is bound; and that however necessary feeling and thought are, as accessory to the action of volition, freedom is found in the will alone.

But if all things were indifferent — if there were no reason why one thing should be chosen rather than another, it is inconceivable that any one act of the will should be better than another; that is to say, there could be no good and no evil. The consciousness of obligation, therefore, either self-imposed, or compelled from without, is a necessary postulate of the knowledge of good and evil. This consciousness of obligation is the discovery of law, and law carries with it the necessary presupposition of a law-giver: and thus we arrive again at God, the author of All Good.

Now, in whatever way we may regard the Mosaic account of the fall of man, — whether we look upon it as a literal and exact account of an event historically true, and hold the garden of Eden to have been an actual garden — the tree of Knowledge, and the tree of Life, actual objects of sense, and so of all else in that simple story; or whether, on the other hand, we look upon the whole of it as pure allegory, the candid and philosophic mind must freely admit the immutable truth underlying it all; and no literary ingenuity, no rugged sagacity, no scientific and technical terminology has ever been able to give better form and expression to the direful truth which it discloses. Man did fail in the beginning, and he fails to-day, to do the very and exact right, by seizing or accepting to his own use that which he knew, and knows, is forbidden him: and so, what the theologian calls sin came into and remains in our humanity as a fact in the consciousness of man, through disobedience.

There is but one conceivable way in which man can be recovered of the disorder caused by, and still kept in our nature by disobedience; and that is, by removing the continuing cause, and putting obedience in the place of disobedience. No mere feeling, however poignant and tender, — no grasp of the intellect, however clear and perspicuous, *as such*, can ever repair the injury, and promote soundness of our moral nature. Holiness, it is true, implies and requires right desires, and honest thinking; but right and honesty are meaningless terms apart from the free activity of the will.

But, again, it must be remembered that heart and mind are not numerically separable from each other, or from the will; but altogether form an essential unity — one indivisible person: so that while rightness implies and requires an uplifting of both heart and mind, the paramount and active agent, in and through the whole self — the ἀρχή — the original and sovereign mode of personal order, is the will.

Now, let it be confessed, that theologians have brought upon themselves much of the distrust and disfavor of which they complain, by assuming a too exact — a too 'He-can,' and 'He-cannot,' 'He-is,' and 'He-is-not,' spirit in dealing with the nature of the Lord of All Power and Might. We cannot, in the nature of things, see the Infinite and Absolute Personality except through human limitations, and it is impious and unscientific to attempt to tear away, or peer through, the limitations which must ever bar us from the Inscrutable: but for all that, let us be careful that we do not fall back upon an equally impious and unscientific Agnosticism, under the specious conviction that we are assuming nothing. While we must admit that we do not and cannot know the Infinite and Absolute, in the sphere of the understanding, rightly directed reflection must teach us that there are the same limitations in any attempt at an ultimate knowledge of the finite and relative; and that, in like sense, we can no more know these than the other. The finite is meaningless apart from the necessary implication of the infinite; and the relative could never have had so much as the name, if the presupposition of the Absolute had not shot through and through what we call things and events. The learned physicist does not know, in a through-and-through sense, what his blow-pipe or scalpel is, — he does not know what water, air, or earth are: he does not even know that they are at all, with the certainty with which he knows that the 'thinker' exists; so that, if the theologian's knowledge runs up into an unknown Infinite Person, his runs down into an Infinite Power, with the difference, if a choice must be made, clearly in favor of the Spiritual Absolute. The theologian does not know all about the One Good; but, through the indisputable consciousness of a sense of obligation, he does know something: and neither does the physicist know all about the materials in his laboratory, and even what he does know, has only a spiritual warrant for its reality. The latter's Agnosticism is wrong end first. He can, with reason, be an agnostic with regard to the Absolute Good, only after he has become an agnostic with regard to the finite 'things,' and that can never be so long as he acknowledges himself and 'things' to be.

The Greeks knew the impossibility of reconciling, in the domain of the mere intellect, the eternal conflict of the limited and the unlimited; and when they asked, How can the one be the many, and the many one? they proposed a problem which has obtruded itself in every effort at philosophic thought since their day, and will continue to confront us until God pleases to lift the veil that we may know Him as He is.

This mystery stares us in the face in whatever direction we may turn; — Matter and Spirit — Identity and Change — Cause and Effect — Life and Death — Good and Evil — Freedom and Necessity — The Infinite and Finite — Being and Non-Being — God and the Universe! No one of all these concepts can be torn away from its correlative, without an utter annihilation of the other; and so, we affirm again, that the visible things of this world are not more certainly known, in the highest sense of knowing, than the Invisible and Ultimate ground of their being.

Theology is Philosophy with a special reference to the nature of God, His relations to man, and man's relations to Him.

Whatever may be construed touching the nature of Deity comes to us, as all construable knowledge comes to us, through the understanding. But the source and ground of the religious element in man is not in the construing power, but in the pre-supposition of the Pure Reason. By virtue of his discovery in himself of a sense of obligation and dependence, it is borne in upon man that there is a Power beyond and above him; and that his well-being is dependent upon such power — beneficent or malignant — as he may look upon it. The religious element is not educated into man, though it may be developed and informed; but when it is absent — if that can ever be — it has been educated out of him. This is abundantly shown by the history of races, and of individuals.

Religious doctrine belongs to the domain of the understanding, while religious convictions belong to the domain of feeling; but the responsibility of developing right convictions and of true desires and aspirations depends upon the free activity of the will. The objective manifestation of the religious element is twofold — looking, from its benevolent side, to the well-being of man, and bearing fruit in alms-deeds, and all manner of benefactions, through the recognition of the paramount obligation to the Good. The other phase becomes articulate in the worship of the All-Good in love and awe of His Infinite Majesty.

It has been said, that as men paint with colors to give an idea of things in strange countries, so God paints with things and peoples and events to give us notions of heavenly and supernatural things. We go with our faces turned down to the ground in search of creeping things, so long as we fail to read the spiritual and divine in what is spread out before us in land and sea, and in the heart of man. He who contrived our hearts in the beginning, and tuned them to the 'level of every day's most quiet need,' gave them, as well, those preternatural strains which lift us into the transcendent world of the Beautiful and True, with a refrain, sweet and mystical, which lifts us higher still into those celestial realms which

> 'God only and good angels know.'

The roar of the sea does not more surely tell of the nearness of the 'multitude of waters,' than this murmur in the heart tells of a supersensible and spiritual world, in us and about us, — a world of spiritual *Realities*, in whose light may be read the manifold riddles of the transitory and seeming. The fact is that any philosophy which has not for its ποῦ στῶ the postulate, man is a spiritual essence manifest in the flesh, is founded upon 'the baseless fabric of a vision,' leads downward and breeds corruption.

There is nothing so common that it does not, if rightly read, lead on to that which is higher; and each stage may be said to be truer and more real than the one which preceded it. All earth and earthly things are types of the Infinite and Eternal: but the heart too often clings to the earth, and earthly things, and calls them real, while, giving only now and then a furtive glance towards the spiritual and abiding, it calls them shadowy and seeming.

And yet the new is always old. When God spoke to Moses in the Mount, and bade him make a sanctuary to the Lord, that He might dwell among his people — when He bade him make the ark, and the mercy seat, and candlesticks, He charged him: 'Look that thou make them after the pattern which I showed thee in the Mount:' or long before, when God made man in the beginning, He formed him after no new or strange device, but said, 'Let us make man in Our image, and after Our likeness'; and 'so God created man in His own image': and thus, throughout the temporal and transitory, we have only likenesses and figures of things higher and truer in the Mount of God. Our life is poor and mean, if we fail to see a

Reality above our work-a-day environment — to peer through the seeming and behold

> "The truth which draws
> Through all things upwards; that a twofold world
> Must go to a perfect cosmos. Natural things
> And spiritual, — who separates these two
> In art, in morals, or in social drift
> Tears up the bond of nature and brings death,
> Paints futile pictures, writes unreal verse,
> Leads vulgar days, deals ignorantly with men,
> Is wrong, in short, at all points. . . .
>   Without the spiritual, observe,
> The natural's impossible; no form,
> No motion! Without sensuous, spiritual
> Is inappreciable; no beauty or power:
> And in this twofold sphere the twofold man
> * * * * * *
> Holds firmly by the natural, to reach
> The spiritual beyond it, — fixes still
> The type with mortal vision, to pierce through,
> With eyes immortal, to the antitype
> Some call the ideal, — better called the real,
> And certain to be called so presently
> When things shall have their names.
> * * * * * *
> Every natural flower that grows on earth
> Implies a flower upon the spiritual side,
> Substantial, archetypal, all aglow
> With blossoming causes, — not so far away,
> That we whose spirit-sense is somewhat cleared
> May catch at something of the bloom and breath.
> * * * * * *
> No lily-muffled hum of summer bee,
> But finds some coupling with the spinning stars,
> No pebble at your feet, but proves a sphere;
> No chaffinch but implies the cherubim.
> * * * * * *
> Earth's crammed with heaven,
> And every common bush afire with God,
> But only he who sees, takes off his shoes."

True as all this is, and clearly as it has been seen by poet and philosopher, all through the ages back to Socrates and Plato, it is a melancholy fact, which every day brings more constantly to light, that there is no holy ground for the empirical philosophy which assumes to guide the spirit of the age. It is so busy with the natural that it fails to see the spiritual, without which the natural is impossible. Its curious gaze is so bent upon the mere mechanism, that it fails to feel the touch of the Infinite Personality which imparts the motion they so much applaud. Lord Bacon, with that mighty spirit of discernment which enabled him to pierce through the outside of things to the reality beyond, says: "As it was aptly said by one of Plato's school, the sense of man resembleth the sun which openeth and revealeth the terrestrial globe, but obscureth the celestial; so doth the sense discover natural things, but shutteth up and darkeneth the divine."

INDEX.

Agnosticism, 198.
Altruism, 320.
Animal world, 95, 152.
Ants, 91.
Apperception, 102.
Architecture, 223.
Art, 220 f.; realism and idealism in, 222 f.
Association, law of, 125.
Atoms, vortex-, 267; 'manufactured articles,' 273.
Automatic action, 87.
Axioms, 177.

Beaver, 93.
Boscovich, 259.
Brain, 24; co-ordination, 26 f.; areas, 27; lesions, 29; electrical stimulations, 31; development, 35; mass, 36.
Browning, Mrs., 339.
Brute creation, the, 95, 152.

Carpenter, Dr., 122.
Categorical Imperative, 228, 323.
Causality, 171; in relation to time and space, 187; inexplicable, 292.
Cell theory, 18.
Certitude, lack of, 1-9.
Change, problem of, 283 f.; influence passing over, 287.
Chætodon Rostratus, 92.
Charcot, Dr., 146.
Chasm between mechanism and consciousness, 69 f.
Choice, 231; freedom of, 314.

Christian faith and evolution, 88.
Cicada, the, 252.
Cognition, 86 f.; no freedom in, 313.
Coleridge, 122, 140.
Color, 64.
Concepts, definite, 102; not *like* objects, 117.
Conscience, functions of, 325 f.; analogy of, with inertia, 326; illustration, 330.
Corti's organ, 49.

Darwin, evolution, 17, 79; physical basis of sensation, 79.
Deduction, 158.
Descartes, anticipated modern psychology, 12 f.
Divine Assistance, 290.
Double consciousness, 144.
Dreaming, 133 f.; dreams within dreams, 141.
Du Bois Reymond, 16; on physical basis of sensation, 80.

Ear, 48 f.
Ego, pure and empirical, 105.
Energy, 193.
Ether, 272.
Evil, origin of, 334.
Evolution and devolution, 95.
Eye, 60 f.

Fechner's law, 41.
Feeling, 199 f.; scheme of, 200; quantity, 204; quality, 205; esoteric and

INDEX.

exoteric, 208; rational, 211; esthetic, 212.
Fichte, 300.
Force, 189; persistent, 192, 198.
Foster, Prof. Michael, 20.

Gases, the kinetic theory of, 262.
Good, the, 227, 316; will of Supreme, how known, 322; the Infinite, 333.
Gravitation, 269.

Hamilton, Sir William, 183.
Hearing, 48.
Hegel, 305.
Helmholtz, 266.
Hobbes, anticipated modern science, 14.
Hume, 171.
Huxley, Professor, cell theory, 19; physical basis of sensation, 77.
Huygens, 15.
Hypnotism, 145 f.
Hypotheticals, 175.

Idealism, 300 f.
Illusions, 218, 254.
Imagination, 126 f.; scheme of, 127.
Induction, 159, 180.
Infinite, nature of concept, 280; conflict of, with finite, 336.
Inhibitory mechanism, 34, 110.
Innate ideas, 163 f.
Instincts, inverse order with intelligence, 94.
Integration and redintegration, 103.

Jelly-specks, 90.

Kant, 301.
Knowledge, immediate, 114.

Law of Excluded Middle, 167; of Contradiction, 166; of Identity, 166; of Sufficient Reason, 168, 175.
Leibnitz, 15, 168, 170.

Light, theory of, 270.
Locke, 169.
Logic, 155 f.
Lotze, on local signs, 41, 119, 302, 309.

Mass, a resistance, 296.
Materialism, 196.
Mathematics, dominates science, 275; contradictions in, 276; surds, 276; cissoid, 278; the infinite, 280.
Matter, gross and sublimated, 250; construction of, 259.
Maudsley, 80.
Maxwell, Professor, 261, 265.
McKendrick, Dr., 26; on physical basis of consciousness, 73.
Mechanics, the foundation of physics, 12 f.; molecular, 266.
Memory, 117; mechanism of, 119.
Metabolism, 21.
Mill, J. S., 181.
Molecules, 263.
Mosso, 139.
Motion, incomprehensible, 295.
Motives, 316.
Music, 214, 223.
Muscles, 32.
Muscular co-ordination, 97.
Mutilations, 26 f.

Nerve-centres, time of action, 30.
Nerves, 22 f.; rapidity of transmission through, 30.
Nervous system, 22.
Newton's laws, 173.

Obedience, law of, 334.
Obligation, sense of, 317.
Occasionalism, 288.
One and the many, the, 10, 105.
Organism, education of, 97.

Painting, 222.
Pasteur, 80.
Penitence, 241.

INDEX.   343

Perception, 113.
'Persistent force,' 189.
Personality, what? 9; psychical factor, 82 f.; ground of action, 188; in relation to energy, 193; unity of, 242; one person and two hypostases, 244; only reality, 251, 298, 311.
Phonograph, 121.
Physical basis of consciousness, 70.
Plato, quoted, 284.
Poetry, 223, 225.
Pre-established harmony, 289.
Pressure spots, 42.
Properties, of bodies, 3, 254.
Protoplasm, 18, 20.
Protozoa, 245.
Psychic factor, 100.
Pure Being, 256.
Pure reason, the, 161 f., 179.

Rational truth, 104.
Reflex action, 23.
Religion, 337.
Romanes, Professor, Rede lecture, 15; motion and sensation, 70.
Rush, Dr., 122.

Schelling, 303.
Schultze, 18.
Schwegler, 307.
Science, relation to older learning, 11; principle of modern, 12.
Sculpture, 222.
Self, an ultimate fact, 9, 111; moral bearing, 313.
Self-consciousness, 107.
Sensation, 86, 110; no liberty in, 313.
Senses, not infallible, 4; specific, 38; touch, 39; taste, 44; smell, 45 f.; hearing, 48; sight, 60.

Skepticism, scope and limit, 1-9; practical and logical limit, 6.
Sleep, 134.
Smell, 45 f.; hearing, 48; sight, 60.
Somnambulism, 143 f.
Sound, 50, 212.
Space, 116, 182, 185, 294 f.
Spencer, Herbert, 193 f.
Structural development, 19.
Sub-consciousness, 87.
Syllogism, the, 155.
Sympathetic system, inhibition, 34.

Tait, Professor, 189, 264.
Taste, 44 f.
Theologians faulted, 335.
'Thing,' what? 254 f., 273.
Thinking, 151; explicit, 162.
Thompson, Sir William, 266.
Threshold value, 40.
Time, 116, 183, 185.
Touch, 38 f.; pressure spots, cold spots, etc., 43.
Tyndall, Professor, sound, 54; physical basis of sensation, 74, 120.

Understanding, the, 149 f.
'Unseen Universe,' quoted, 250.

Vision, 216; illusions, 218.
Vivisection, 26, 29.
Voice, the human, 57.

Weber's law, 41.
Will, the, 229 f.; conscious volition, 230; liberty restricted, 231; inhibitory function, 232; moral aspect, 240.
Wordsworth, 217.
Wundt, 15.

www.ingramcontent.com/pod-product-compliance
Lightning Source LLC
Chambersburg PA
CBHW020241240426

43672CB00006B/600